WETLAND BIRDS AND SEABIRDS

FROM OSPREYS TO PUFFINS

This edition published in 2010 by:

CHARTWELL BOOKS, INC.
A Division of BOOK SALES, INC.
276 Fifth Avenue Suite 206
New York, New York 10001
U.S.A.

ISBN 13: 978-0-7858-2600-2
ISBN 10: 0-7858-2600-9

Printed and bound in China by Midas Printing Ltd.

A Marshall Edition
Conceived, edited, and designed by Marshall Editions
The Old Brewery
6 Blundell Street
London N7 9BH, UK
www.quarto.com

10 9 8 7 6 5 4 3 2 1

Art Director Ivo Marloh
Editorial Elise See Tai
Production Nikki Ingram

WETLAND BIRDS AND SEABIRDS

FROM OSPREYS TO PUFFINS

Editorial Consultant

Rob Hume

CHARTWELL
BOOKS, INC.

Killdeer
Widespread throughout the Americas, this plover occasionally visits Europe and Hawaii.

Common Murre
This bird spends most of its time at sea, coming to land only to breed in large colonies on cliff ledges.

White Stork
These storks sometimes nest on cliffs and ledges but most often make their home on buildings.

Royal Tern
This tern winters on the southern coasts of the U.S.A.

Wilson's Storm-petrel
Oceanites oceanicus
One of the most abundant
bird species, this seabird
regularly follows in the
wake of ships.

CONTENTS

Arctic Loon
Gavia arctica
When feeding, this
bird dives to depths of
10–20 ft (3–6 m) for
about 45 seconds.

INTRODUCTION

Wetlands include areas with permanent, shallow, or temporary water, both salt and fresh: From the open sea to lakes, rivers, marshes, and wash lands that flood periodically. Many shallow wetland habitats naturally dry out, through waterlogged thickets to woodland. In a natural landscape, they would be replaced by new wetlands, as rivers change course, sandbanks back up new coastal marshes, and sea levels change relative to the land. In the modern world, however, many wetlands are contained or drained and the scope for natural development is much reduced. Many wetland birds therefore rely increasingly on artificial wetlands, such as reservoirs and flooded gravel pits, or on conservation management that maintains fragile reedbeds and swamps.

Birds that occupy such areas are many and varied. At sea, there are petrels, albatrosses, shearwaters, cormorants, and gulls. Feeding on and under freshwater are divers, grebes, ducks, geese, and swans. Marshes have ducks, geese, herons, and egrets, and various birds of prey that specialize in fish or other wetland animals as prey. Gulls and terns feed and may also breed on wetland areas. Reed swamps, with their tall, dense columns of reed stems, are demanding habitats that are home to bitterns and warblers adapted to grasp

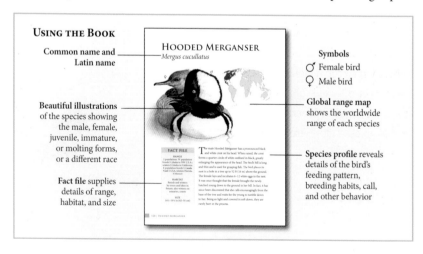

USING THE BOOK

Common name and Latin name

Beautiful illustrations of the species showing the male, female, juvenile, immature, or molting forms, or a different race

Fact file supplies details of range, habitat, and size

HOODED MERGANSER
Mergus cucullatus

Symbols

♂ Female bird
♀ Male bird

Global range map shows the worldwide range of each species

Species profile reveals details of the bird's feeding pattern, breeding habits, call, and other behavior

upright stems and build nests slung between stalks. Crakes and rails creep inconspicuously beneath, and kingfishers hunt at the edge of open spaces with clear, shallow water.

Ringed Plover
Charadrius hiaticula
If a potential predator nears its nest, this bird pretends to be injured, using its broken-wing display to distract the predator's attention.

AQUATIC HARVESTS

Wetlands and seas offer rich pickings for many birds. The open sea, for example, varies from near-sterile areas to hugely productive regions with great numbers of birds, and tidal wetlands benefit from the twice-daily tide that brings fresh nutrients and food. Estuarine environments are among the richest on Earth, comparable with rain forests and more productive than the best farmland. As the mud, sand, silt, and rocks are twice daily inundated by seawater and exposed to drying air, the creatures that live there must be well suited to a demanding lifestyle, and birds are well adapted to take advantage of the worms, crustaceans, and mollusks. Birds rarely survive year-round in the same place, however, and all seabirds must come to land to breed. From fall to spring, temperate and tropical estuaries provide refuges for millions of birds that move north to breed in Arctic regions, but offer relatively few opportunities for birds to nest around them. Wetlands in general are home to some of the world's greatest globe-trotters in the bird world.

Rob Hume

Dovekie (Little Auk)
Alle alle
The Dovekie's breeding colonies may contain several million pairs.

Christmas Island Frigatebird
Fregata andrewsi
Male frigatebirds display to
females by inflating their
scarlet throat pouch
and warbling.

Manx Shearwater
Puffinus puffinus
This bird was once a rare visitor
to North America but has now
established small populations on
the northeast Atlantic coast.

Atlantic Puffin
Fratercula arctica
The beak of this bird is
used to attract mates,
and its razor-sharp
edges help it hold up
to a dozen fish at once.

Northern Fulmar
Fulmarus glacialis
A relative of the albatrosses, this
bird breeds off Alaska and Canada.

Birds
of the
Open Seas

The great wanderers of the open ocean range
from huge albatrosses up to 11 ft (3.5 m)
across the wings, to storm-petrels of just a few
inches long. In-between are larger petrels and
shearwaters that exploit air currents over waves to
fly long distances. Petrels and shearwaters mainly
visit their nests at night because their short,
weak legs make them vulnerable to land-based
predators. Many locate their nests using smell
and by homing in on the voice of their mates
in a remarkable nocturnal cacophony of calls.

    Frigatebirds cannot settle on the sea. Instead,
these magnificent low-energy fliers glide great
distances, stealing fish from other birds. In
contrast, gannets and boobies dive from great
heights deep into the water to chase their prey.
In the northern hemisphere, auks, such as
puffins, guillemots, and razorbills, also dive for
fish and nest in huge colonies. In the southern
hemisphere, the auks' role is performed by
penguins, flightless birds insulated by
thick layers of fat, dense feathers,
and flipperlike wings.

Northern Gannet
Sula bassana
Diving from heights of up to
100 ft (30 m), this bird enters
the water at great speed to
hunt mackerel and herring.

Imperial Shag
Phalacrocorax atriceps
This cormorant hunts
for fish at depths of at
least 80 ft (25 m).

KING PENGUIN
Aptenodytes patagonicus

fully grown chick

The King Penguin raises 2 chicks every 3 years. A single egg laid in November is incubated by both parents for 55 days. By June, the chick weighs 80 percent of its adult weight. It survives the winter on very little food, banded together with other chicks in crèches to resist the cold. Feeding starts again in September and the chick fledges 2–3 months later. Meanwhile, the parent birds have to molt, and cannot lay again until February (late summer). By winter, the chicks are still very small and many die during the next few months. Those that survive, fledge the following January. Parents following this timetable cannot breed again until the following summer. In some colonies, it seems that most adults breed only once every 2 years.

FACT FILE

RANGE
Subantarctic, Falkland Islands

HABITAT
Oceanic; breeds on coasts

SIZE
37½ in (95 cm)

ADÉLIE PENGUIN
Pygoscelis adeliae

well-grown chick

The most southerly of the penguins, apart from the Emperor, the Adélie nests in large colonies on islands and headlands around much of the Antarctic coast. Suitable breeding sites are limited by the availability of ground that becomes bare of snow in summer, near water that becomes free of ice; the breeding adults must have ready access to open water to provide food for their chicks. Experienced breeders return to the same sites year after year and usually to the same mates. The Antarctic summer is very short, so they have little time to breed. The eggs are laid in late November (mid-summer) and incubated first by the male and then by the female for a total of 35 days. After 3–4 weeks in the nest, the dark, woolly chicks band together in crèches while the adults gather food. In another 4 weeks the chicks are fledged, going to sea in February (late summer) when they are only three-quarters of the adult weight.

FACT FILE

RANGE
Antarctic

HABITAT
Oceanic; breeds on
rocky coasts

SIZE
28 in (71 cm)

JACKASS PENGUIN
Spheniscus demersus

juv

There were once a million or more Jackass Penguins, but the population has dropped to 150,000 in the last 20 years. The birds were badly affected by the oil pollution that first hit their breeding and feeding areas in 1967, when the Suez Canal was closed and oil tankers were diverted around southern Africa. Today, they are under threat from increased commercial fishing for the pilchards that once made up half their diet. Although the Jackass Penguin feeds in the cool Benguela Current, it breeds on dry land under the hot African sun. It avoids overheating by nesting in burrows or under rocks, and by being active on land only at night.

FACT FILE

RANGE
Coasts of S Africa

HABITAT
Cool coastal seas;
breeds on coast

SIZE
27½ in (70 cm)

WANDERING ALBATROSS
Diomedea exulans

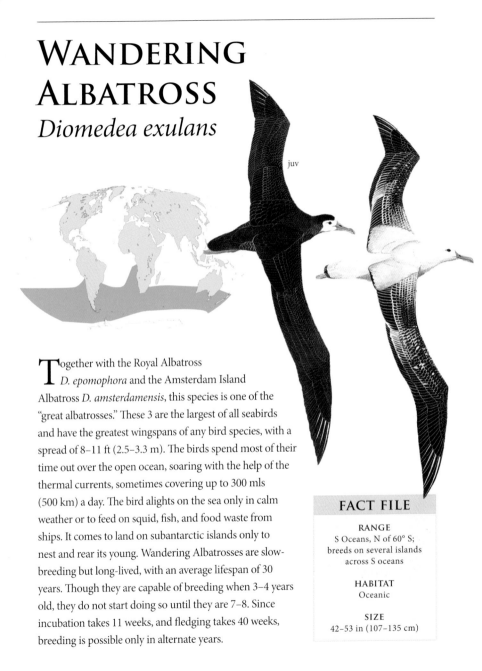

juv

Together with the Royal Albatross *D. epomophora* and the Amsterdam Island Albatross *D. amsterdamensis*, this species is one of the "great albatrosses." These 3 are the largest of all seabirds and have the greatest wingspans of any bird species, with a spread of 8–11 ft (2.5–3.3 m). The birds spend most of their time out over the open ocean, soaring with the help of the thermal currents, sometimes covering up to 300 mls (500 km) a day. The bird alights on the sea only in calm weather or to feed on squid, fish, and food waste from ships. It comes to land on subantarctic islands only to nest and rear its young. Wandering Albatrosses are slow-breeding but long-lived, with an average lifespan of 30 years. Though they are capable of breeding when 3–4 years old, they do not start doing so until they are 7–8. Since incubation takes 11 weeks, and fledging takes 40 weeks, breeding is possible only in alternate years.

FACT FILE

RANGE
S Oceans, N of 60° S;
breeds on several islands
across S oceans

HABITAT
Oceanic

SIZE
42–53 in (107–135 cm)

SOOTY ALBATROSS
Phoebetria fusca

FACT FILE

RANGE
Atlantic and Indian oceans;
breeds on several islands in
this region

HABITAT
Oceanic

SIZE
33–35 in (84–89 cm)

The Sooty Albatross and its close relative, the Light-mantled Sooty Albatross *P. palpebrata*, differ from other members of their family in their proportionality longer, narrower wings and tapering tail and more solitary nesting habits. Their flight is graceful and agile and their displays include synchronized flights up and down in front of the colony by courting and established pairs. They breed on cliffs or at sites inland where there is sufficient slope in front of the nests to enable them to take off easily. They start breeding when they are 11–13 years old, after a 3–4 year courtship, and young are reared only in alternate years.

BLACK-BROWED ALBATROSS
Diomedea melanophris

adult feeding
chick at nest

One of the most numerous of the albatrosses, this species nests alongside the similar Gray-headed Albatross *D. chrysostoma*, often gathering in great colonies. The nest is a barrel-shaped pile of mud and grass, about 24 in (60 cm) high. The male arrives at the colony a week before the female, who stays for 1 day for mating, goes to sea for another 10 days, and finally returns 2 days before laying. The species' main food is krill, which may explain why it can nest every year, while the Gray-headed Albatross, which feeds on nutritionally poor squid, breeds only in alternate years.

FACT FILE

RANGE
S oceans, from 65° to 10°
S; breeds on several islands
across S oceans

HABITAT
Oceanic

SIZE
32½–36½ in (83–93 cm)

SOUTHERN GIANT PETREL
Macronectes giganteus

juv

white
phase

FACT FILE

RANGE
S oceans, N to 10° S; breeds
on various islands and on
Antarctic coasts

HABITAT
Oceanic

SIZE
34–39 in (86–99 cm)

Popularly called Stinkers because of their rank smell and their habit of feeding on carrion, the Southern and Northern Giant Petrel *M. halli* are very similar species. The largest of the petrels, these birds rival some of the albatrosses in size. They are agile on land and feed on land as well as at sea. They use their massive bills to rip into the flesh of dead animals, such as seals and whales. The nest is a low saucer of pebbles or soil and vegetation. Both sitting adults and nestlings defend themselves with the common petrel habit of spitting oil. The white phase of this species is spread through the breeding population, accounting for 15 percent of birds in places.

WILSON'S STORM-PETREL
Oceanites oceanicus

Arguably the most abundant seabird in the world, Wilson's Storm-petrel regularly gathers in large flocks to feed and follows in the wake of ships. This small bird's diet consists largely of crustaceans, which it catches by plucking them from the surface, pattering its feet on the water to "anchor" itself against the breeze. The egg, laid in a burrow, is one quarter of the adult's weight. At its peak, the nestling reaches twice the adult weight; it fledges at 1½ times the adult weight.

FACT FILE

RANGE
Breeds from Tierra del Fuego and the Falkland Islands to the coasts of Antarctica; migrates to N Indian Ocean and N Atlantic Ocean after breeding season

HABITAT
Oceanic

SIZE
6–7½ in (15–19 cm)

NORTHERN FULMAR
Fulmarus glacialis

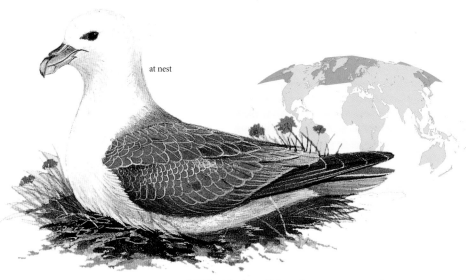

at nest

FACT FILE

RANGE
North Pacific and
Atlantic oceans

HABITAT
Oceanic; breeds
on coasts

SIZE
18–19½ in (45–50 cm)

The Northern Fulmar has grown in numbers over the last 100 years, either because of the increased quantities of offal discarded first by whalers and now by fishing boats, or through changes in ocean water temperatures. St. Kilda once had the only colony in the British Isles, but colonies are now found on all cliff coasts and the total British and Irish population is now more than 350,000 pairs. The nests are sited on sea cliffs but sometimes farther inland on cliffs, walls, and buildings. The birds visit the sites for most of the year, performing soaring displays and uttering cackling calls. Of the 2 plumage forms, the pale phase is the more numerous in the Arctic, the dark phase at lower latitudes, although the detailed picture is complex and a range of intermediates also occurs.

CORY'S SHEARWATER
Calonectris diomedea

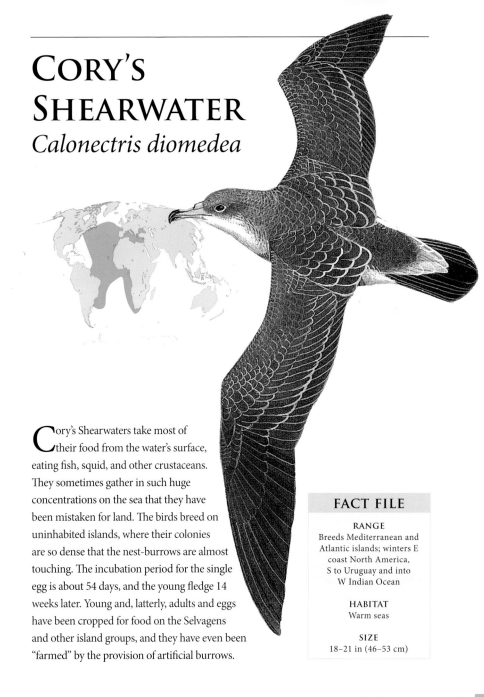

Cory's Shearwaters take most of their food from the water's surface, eating fish, squid, and other crustaceans. They sometimes gather in such huge concentrations on the sea that they have been mistaken for land. The birds breed on uninhabited islands, where their colonies are so dense that the nest-burrows are almost touching. The incubation period for the single egg is about 54 days, and the young fledge 14 weeks later. Young and, latterly, adults and eggs have been cropped for food on the Selvagens and other island groups, and they have even been "farmed" by the provision of artificial burrows.

FACT FILE

RANGE
Breeds Mediterranean and Atlantic islands; winters E coast North America, S to Uruguay and into W Indian Ocean

HABITAT
Warm seas

SIZE
18–21 in (46–53 cm)

Manx Shearwater
Puffinus puffinus

race *mauretanicus*
"Balearic Shearwater"

The contrast between the dark upperparts and light underparts of the Manx Shearwater is much less distinct in the browner Balearic Shearwater *P. mauretanicus*, a species that breeds in the western Mediterranean. Manx Shearwaters nest in colonies, usually on isolated islands. They arrive at the colonies as early as February and choose mainly the darkest nights to visit their burrows as a defense against attacks from Great Black-backed Gulls *Larus marinus*. The pairs call to one another with a variety of loud, raucous screams and howls. The chicks fledge at 70 days, but are deserted by their parents 1–2 weeks earlier.

FACT FILE

RANGE
Breeds N Atlantic and Mediterranean islands; migrates S as far as South America after breeding

HABITAT
Offshore seas above continental shelf

SIZE
12–15 in (30–38 cm)

RED-FOOTED BOOBY
Sula sula

white phase

white phase
Galapagos form

white-tailed
brown phase

The Red-footed Booby is probably the most numerous member of its family. There are several color phases across its huge range, but all have conspicuous red feet. Unlike most other members of the family, the species nests in trees. This habit may have partly protected it from human disturbance, helping to account for its comparative abundance today. In their relatively non-seasonal environment, the birds breed opportunistically, waiting until there is an ample food supply. However, it is common for the supply to fail at a later stage, causing heavy mortality among the young. The surviving juveniles disperse widely and may travel thousands of miles from their birthplace. They can be distinguished from the adults by their blackish-brown bills, purplish facial skin, and yellowish legs.

FACT FILE

RANGE
All tropical oceans

HABITAT
Open sea

SIZE
28–31 in (71–79 cm)

BROWN BOOBY
Sula leucogaster

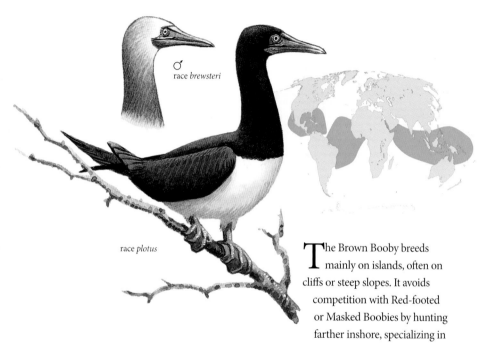

♂
race *brewsteri*

race *plotus*

The Brown Booby breeds mainly on islands, often on cliffs or steep slopes. It avoids competition with Red-footed or Masked Boobies by hunting farther inshore, specializing in low, slanting dives to catch fish and squid. It spends much time in the air and performs some of its courtship rituals in flight. The breeding colonies are smaller and more scattered than those of the Red-footed and Masked Boobies. As with the Masked Booby, older chicks kill their siblings. On many islands, the birds breed annually, but on Ascension Island in the Atlantic, they breed at less than yearly intervals, triggered by favorable changes in the food supply. *S. l. plotus* breeds in and around the Indian Ocean, Australia, and some Pacific islands; *S. l. brewsteri* nests in the eastern Pacific and along the western coast of Mexico.

FACT FILE

RANGE
Coastlines of all
tropical oceans

HABITAT
Open sea

SIZE
29½–31½ in (75–80 cm)

NORTHERN GANNET
Sula bassana

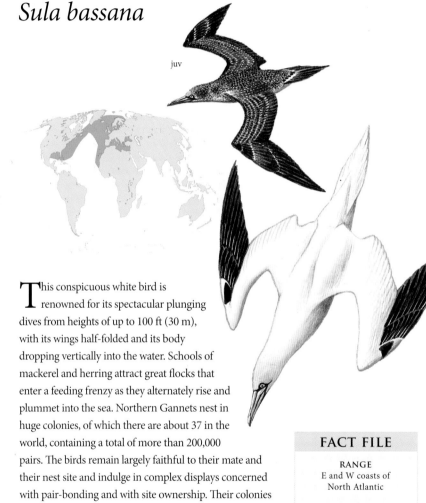

juv

This conspicuous white bird is renowned for its spectacular plunging dives from heights of up to 100 ft (30 m), with its wings half-folded and its body dropping vertically into the water. Schools of mackerel and herring attract great flocks that enter a feeding frenzy as they alternately rise and plummet into the sea. Northern Gannets nest in huge colonies, of which there are about 37 in the world, containing a total of more than 200,000 pairs. The birds remain largely faithful to their mate and their nest site and indulge in complex displays concerned with pair-bonding and with site ownership. Their colonies are intensely active and noisy. Because the peak food supply is both seasonal and dependable, egg-laying takes place on a consistently similar date every year. With an assured food supply, the single chick has a high chance of survival.

FACT FILE

RANGE
E and W coasts of
North Atlantic

HABITAT
Marine offshore

SIZE
35½ in (90 cm)

Red-legged Cormorant
Phalacrocorax gaimardi

FACT FILE

RANGE
SW South America

HABITAT
Marine coasts

SIZE
28–30 in (71–76 cm)

This strikingly handsome bird is unusual among the cormorants for its relatively solitary habits, unexpectedly chirpy voice, and its restricted distribution on the Pacific coast of South America. It occurs on almost all of the Peruvian Guano Islands, but often sparsely, nesting alone or in small groups in cavities on sea cliffs. Only occasionally do the birds cluster densely enough to be termed colonial. Similarly, when foraging, they often fly singly for long distances and hunt alone or in pairs though feeding groups sometimes occur. The birds build large nests, which include items, such as the cases of tube-dwelling worms and fronds of gelatinous seaweed, which glue the structure to the rock.

SHAG
Phalacrocorax aristotelis

non-breeding

The adult Shag is an attractive bird with blackish-green plumage, green eyes, and a long, wispy, forward-curving crest in the breeding season. Although usually a quiet bird like other cormorants, it has a sonorous grunting call, often accompanied by throat-clicking. It prefers clear, cold seas, avoiding estuaries and inland waters, and will travel up to 7 mls (11 km) from its colony in order to feed. Typically, Shag colonies are neither large nor dense, although some reach a total of more than 1,000 pairs. After breeding, the birds tend to disperse along the coast, but northern populations, such as those from northern Norway, may migrate more than 600 mls (1,000 km) south. There is evidence that some breeding colonies have their own specific wintering areas.

FACT FILE

RANGE
Coastal W and S Europe and N Africa

HABITAT
Marine coasts

SIZE
25½–31½ in (65–80 cm)

IMPERIAL SHAG
Phalacrocorax atriceps

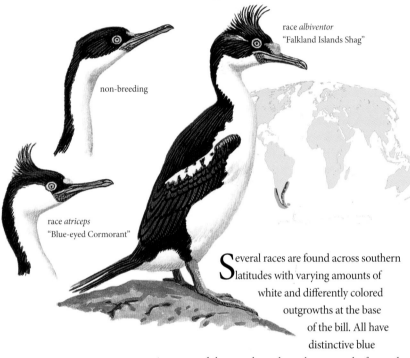

race *albiventor*
"Falkland Islands Shag"

non-breeding

race *atriceps*
"Blue-eyed Cormorant"

Several races are found across southern latitudes with varying amounts of white and differently colored outgrowths at the base of the bill. All have distinctive blue rings around the eyes throughout the year and a forward-curving, wispy crest in the breeding season. Two races are illustrated: the Falkland Islands Shag or King Cormorant *P. a. albiventor* and the Blue-eyed Cormorant *P. a. atriceps* of mainland South America. The Falklands Islands Shag is an isolated race and the only cormorant that occurs on these South Atlantic islands. It forms huge colonies on flattish rocks and may also nest among penguins and albatrosses. The Blue-eyed Cormorant, by contrast, is widespread along the coast of southern Chile and Argentina.

FACT FILE

RANGE
S South America,
subantarctic islands,
Antarctic peninsula

HABITAT
Marine coasts

SIZE
28 in (72 cm)

CHRISTMAS ISLAND FRIGATEBIRD

Fregata andrewsi

B reeding only on Christmas Island in the Indian Ocean, this species is reduced to fewer than 1,000 pairs. They are now protected, after centuries of human persecution. The Frigatebird occasionally causes the death of another rare endemic bird, Abbott's Booby *Sula abbotti*, by hounding it for fish and forcing it to crash down into the jungle canopy. Like other frigatebirds, the males display in small groups to passing females by presenting their inflated scarlet throat pouch, trembling their outspread wings, and producing a warbling call. If a female is attracted by a male's courtship display, she descends alongside him. After mating, the pair build a nest of twigs on the display site.

FACT FILE

RANGE
Christmas Island and surrounding seas

HABITAT
Oceanic

SIZE
35–40 in (89–102 cm)

GREAT FRIGATEBIRD
Fregata minor

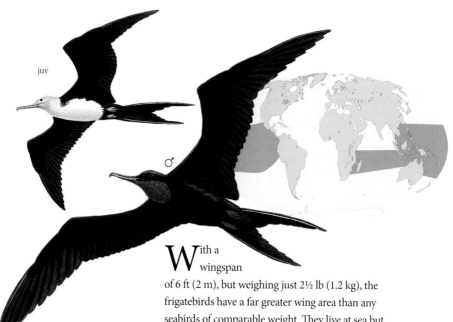

juv

♂

With a wingspan of 6 ft (2 m), but weighing just 2½ lb (1.2 kg), the frigatebirds have a far greater wing area than any seabirds of comparable weight. They live at sea but seldom alight on water, feeding on the wing by snatching surface prey. Some steal fish from other seabirds. Though they lack waterproof plumage, they range hundreds of miles from land. Thousands of Great Frigatebirds breed in treetop colonies on oceanic islands. The breeding cycle is one of the longest of any seabird—the chick requiring 6 months to fledge and a further 6 months of support from its parents, which can breed only once every 2 years. "Rest" years reduce productivity, and aggressive non-breeding males cause such disturbance in colonies that many eggs and young are lost.

FACT FILE

RANGE
Tropical and subtropical Indian and Pacific oceans, and off Brazil

HABITAT
Open sea

SIZE
34–39 in (86–100 cm)

COMMON MURRE (GUILLEMOT)

Uria aalge

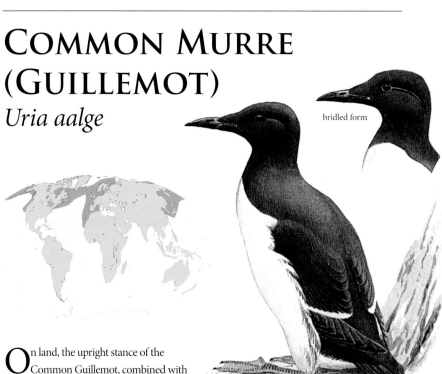

bridled form

O n land, the upright stance of the
Common Guillemot, combined with
its dark head and upperparts and white
underparts, give the bird the appearance
of a small penguin. The bridled form
is a genetic variation that occurs only in the North Atlantic,
often in northern populations. Common Guillemots breed in
large colonies on narrow cliff ledges. Each pair's single egg is
markedly pointed so that it will roll in a circle if pushed and
will not fall off the ledge. The chicks leave the nest site when
they are only about 3 weeks old. Their wing covert feathers
grow first, giving them just enough power to half-glide, half-
fly from the ledge down to the water. The chicks then swim
away out to sea, usually with the male parents. This early
departure from the ledges means that the parents can take
their young to the food supply instead of continuing to bring
the food to them from as far as 50 mls (80 km) away.

FACT FILE

RANGE
Circumpolar Eurasia and
North America

HABITAT
Open sea; breeds
on coastal cliffs

SIZE
16–17 in (40–43 cm)

DOVEKIE (LITTLE AUK)
Alle alle

The Dovekie is a small, stubby seabird, with little or no neck and short wings. Its principal food is plankton, small crustaceans, and other marine animals. During the breeding season, it brings food back to the nest in its crop, which, when full, produces a distinct bulge in the throat. Breeding colonies may contain several million pairs of Dovekies. Their nests are placed deep among the fallen rocks of coastal scree. Colonies are constantly patrolled by the species' main predators, Glaucous Gulls, which wait for their chance to snatch unwary birds as they enter or leave their nest-holes.

FACT FILE

RANGE
North Atlantic and adjacent Arctic

HABITAT
Breeds on coastal scree and in rock crevices; otherwise on open sea

SIZE
8–10 in (20–25 cm)

BLACK GUILLEMOT
Cepphus grylle

non-breeding

In contrast
to its pale,
mottled winter plumage, the
breeding dress of the bird is a bold smoky black apart
from a broad white patch on the upper side of the wing
and white on the inner half of the underwing. The bird also
has bright red legs and feet that show up distinctly in clear
water and on land. It utters a loud and melodious whistling
call, both in flight and on the water, and as it calls, it reveals
the bright red lining of its mouth. During courtship,
one bird circles closely around its mate, producing a
softer version of the whistle at frequent intervals. Black
Guillemots breed alone or in small colonies of up to 100
pairs. They may nest under boulders, in rock crevices,
or in caves, laying their eggs on the bare earth or rock.

FACT FILE

RANGE
North Atlantic and
adjacent Arctic

HABITAT
Breeds on coasts; winters
mainly in coastal waters

SIZE
12–14 in (30–36 cm)

MARBLED MURRELET
Brachyramphus marmoratus

non-breeding

FACT FILE

RANGE
N Pacific, S to Japan
and California

HABITAT
Inshore marine; breeds in
woodland and tundra near
the coast

SIZE
9½–10 in (24–25 cm)

The Marbled Murrelet takes its name from the pale edgings to its feathers that give the bird a mottled appearance, especially in the breeding season. Although it is an abundant species, with hundreds of thousands occurring off the coasts of Alaska, its breeding habits are virtually unknown. Only a few solitary nests have ever been found. Two of these were several miles inland in forested areas, placed in the fork of a large branch and in a tree hollow about 100 ft (30 m) above the ground. Two other nests were in hollows in the open tundra of small subarctic islands.

ATLANTIC PUFFIN
Fratercula arctica

non-breeding

The most distinctive feature of the Atlantic Puffin is its large red, yellow, and gray-blue bill, shaped rather like the bill of a parrot. The coloring of this extraordinary appendage is particularly bright during the breeding season and helps to attract mates. The depth of the bill, and especially its sharp edges, enable the bird to catch and hold fish. A dozen or more can be held at one time and, as each fish is caught, it is held between the tongue and the upper mandible, freeing the lower mandible for further catches. Atlantic Puffins also use their beaks to excavate nesting burrows, which can be several feet long, in grassy clifftops.

FACT FILE

RANGE
N Atlantic and
adjacent Arctic

HABITAT
Open sea; breeds on coasts

SIZE
11–12 in (28–30 cm)

Pied-billed Grebe
Podilymbus podiceps
Many Pied-billed Grebes
will stay in the same
stretch of water during
summer and winter.

Great White Pelican
Pelecanus onocrotalus
This bird collects
fish in its throat pouch.

Horned (Slavonian) Grebe
Podiceps auritus
This bird makes its nest from
aquatic vegetation.

Great Cormorant
Phalacrocorax carbo
The Great Cormorant dives
underwater to catch prey.

PRIMITIVE WETLAND SPECIALISTS

Brown Pelican
Pelecanus occidentalis
This pelican makes dives into the water from great heights.

Divers and grebes are well-adapted waterbirds. Their legs are set so far back that they are barely able to walk, but they are strong and adapted for swimming. Their feet are not webbed, but their toes have broad lobes that fold back on the forward stroke and spread wide on the thrusting back stroke, to give forward propulsion.

Divers have well-developed, wailing calls while grebes rely on elaborate head adornments and ritualized movements in their displays.

Pelicans, cormorants, and darters are related to gannets, and all four toes are joined by webs, which provide a different means of propulsion. Pelicans are big, heavy birds but are magnificent in the air. They feed by diving or while swimming, often in flocks, using a large, soft pouch that expands beneath the bill to scoop up fish. Darters, or snake birds, have large bodies and long wings, but very slender heads and slim, daggerlike bills— they catch fish while diving underwater.

Australian Pelican
Pelecanus conspicillatus
This sociable bird feeds mainly on fish, but also eats other aquatic animals.

Red-throated Loon (Diver)

Gavia stellata

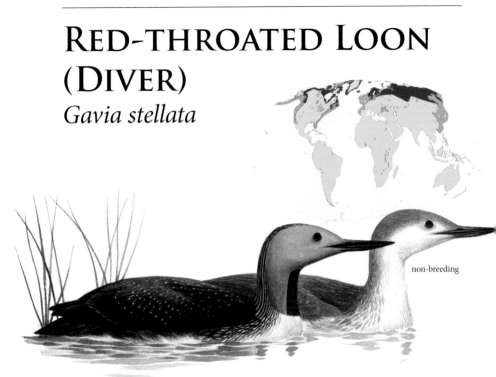

non-breeding

The Red-throated Loon is a slender, graceful bird, with a slightly upturned bill and a rusty-red throat patch. Unlike other members of its family, which require a long run across the water to become airborne, the Red-throated Loon can take off easily from small bodies of water. This allows it to nest beside pools only 33 ft (10 m) across in its breeding grounds on the Arctic tundra and on temperate moorland. Such pools seldom contain enough fish and the birds fly several miles to larger lakes or to the sea to obtain food for their young. The female usually lays 2 eggs and they are incubated by both parents for about 28 days. After hatching, competition between the young for the food brought by the parents frequently leads to the death of the weaker chick.

FACT FILE

RANGE
Circumpolar Arctic S to temperate latitudes

HABITAT
Breeds on freshwater lakes and pools; winters mainly on coastal waters

SIZE
21–27 in (53–69 cm)

Arctic Loon
Gavia arctica

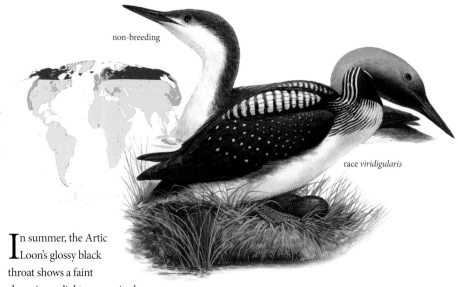

non-breeding

race *viridigularis*

In summer, the Artic Loon's glossy black throat shows a faint sheen in sunlight—green in the Siberian race G. a. *viridigularis*, purple in the European race G. a. *arctica*. Those from North America and coastal Siberia are separated into a different species, the Pacific Loon G. *pacifica*, and also shine purple. Arctic Loons breed beside large lakes where they find most of their food. When feeding, they dive for about 45 seconds, probably reaching depths of 10–20 ft (3–6 m). They occasionally use their wings under the water to assist their webbed feet. The nest is a low mound of aquatic vegetation next to the water. The normal clutch is 2 and incubation lasts for about 28 days. Fledging takes a further 12–13 weeks, by which time the young birds are beginning to catch their own fish.

FACT FILE

RANGE
Circumpolar Arctic and N temperate latitudes; winters to S

HABITAT
Breeds on large, deep lakes; winters on coastal waters

SIZE
23–29 in (58–73 cm)

COMMON LOON
Gavia immer

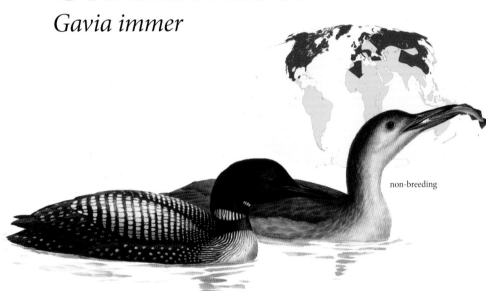

non-breeding

This is a bigger, heavier, larger-billed bird than other divers. In the breeding season, its upperparts are patterned all over with bold white barring and 2 white necklaces mark its black head and neck. Its calls, used to advertise and defend the breeding territory, are among the loudest of any bird and consist of a variety of yodeling, howling, and wailing cries that can carry for long distances across the water. At their most intense, these calls sound like manic laughter. It has been suggested that "loon," originally came from the old Norse name *lomr. Lomr* may mean to moan, in which case it refers to the birds' wailing calls, but some sources suggest it means lame or clumsy and refers to the birds' awkward gait on land. A loon's feet are set far back on its body so it usually humps itself along on its breast and belly. It builds its nest close to the water's edge so that it has the minimum distance to move on land.

FACT FILE

RANGE
N North America, Greenland and Iceland; winters to S

HABITAT
Breeds on large, deep lakes; winters on coastal waters

SIZE
27–36 in (69–91 cm)

PIED-BILLED GREBE
Podilymbus podiceps

non-breeding

The short, stocky body of the Pied-billed Grebe contrasts markedly with the bird's large head and prominent bill. The bill is not unlike a chicken's, nearly half as deep as it is long. It is whitish in color with a black band at the midpoint, making it the most conspicuous feature on this otherwise rather dull dark gray-brown bird. Many Pied-billed Grebes are sedentary, spending summer and winter on the same stretch of water. They sometimes form loose flocks in winter but break up into breeding pairs and establish territories in the spring. In the northern parts of the range, the lakes and pools freeze over in winter. Birds that breed there either migrate south to find suitable winter quarters or move to brackish, ice-free lagoons on the coast.

FACT FILE

RANGE
S South America N
to S Canada

HABITAT
Shallow standing or slow-
moving freshwater

SIZE
12–15 in (31–38 cm)

GREAT CRESTED GREBE

Podiceps cristatus

non-breeding

FACT FILE

RANGE
Temperate Eurasia, E and South Africa, Australia and New Zealand

HABITAT
Shallow, standing or slow-moving freshwater with emergent vegetation; moves to estuaries and coastal lagoons in winter

SIZE
18–20 in (46–51 cm)

The Great Crested Grebe is renowned for its elaborate courtship rituals. Both sexes have a double black head crest and chestnut and black tippets (ear ruffs) and, during courtship, they erect these feathers, which gives their heads a markedly triangular outline. The pair indulge in a "weed-dance," during which pieces of waterweed are held in the bill as the 2 birds come together, breast to breast, and rear up out of the water, swinging their heads from side to side. The dance is both elegant and stately. Like other grebes, the Great Crested Grebe feeds on fish, insects, crustaceans, and mollusks. When the young hatch, the parents feed them small fish and also a supply of small feathers. Adults, too, will regularly swallow feathers molted from their belly and flanks. It is believed that the feathers bind material together in the gut, helping the birds regurgitate waste pellets. Regular regurgitation probably helps to clear the gut of harmful parasites.

WESTERN GREBE, CLARK'S GREBE

Aechmophorus occidentalis, A. clarkii

Clark's Grebe

Western Grebe

Until recently Clark's Grebe (also known as the Mexican Grebe) was considered a race of the Western Grebe but it is now regarded as a separate species. The birds share the same general range and habitat, although Clark's Grebe is much less common to the north. Hybrids between the 2 species are occasionally reported. Western Grebes perform a series of elaborate courtship rituals. During their "rushing" ceremony, for example, male and female rise up out of the water together. Holding themselves erect, with only their legs under the surface, they rush along side by side for about 22–33 yd (20–30 m) before sinking back into the water. Western Grebes are fairly gregarious, with several birds breeding on the same water. Flocks of many hundreds or even thousands gather at migration stopovers and in the wintering areas.

FACT FILE

RANGE
W North America, from
S Canada to Mexico

HABITAT
Breeds on freshwater lakes;
winters mainly on the sea

SIZE
22–29 in (56–74 cm)

HORNED (SLAVONIAN) GREBE

Podiceps auritus

non-breeding

FACT FILE

RANGE
Circumpolar N North
America and Eurasia

HABITAT
Breeds on both small and
large waters; winters on large
lakes and sheltered coasts

SIZE
12¼–15 in (31–38 cm)

The courtship behavior of the Horned Grebe, like that of the Great Crested Grebe, is dramatic and elaborate. To enhance their appearance, the birds erect the golden-chestnut crests that run back and upward from each eye and the black tippets, or ear ruffs, under their eyes. In one ritual, however, the display feathers are smoothed down as one bird rears up out of the water and bends its head and neck downward. As it does so, it bears a striking resemblance to a penguin and this particular display has been called the "ghostly penguin dance." Horned Grebes gather a pile of aquatic vegetation for their nest. Sometimes they anchor the pile to plants in the water, but often the nest rests on a small rock at or just below the water level.

GREAT WHITE PELICAN
Pelecanus onocrotalus

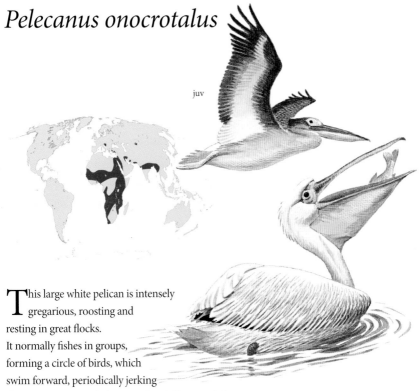

juv

This large white pelican is intensely gregarious, roosting and resting in great flocks. It normally fishes in groups, forming a circle of birds, which swim forward, periodically jerking their wings open and plunging their bills toward the center of the circle. In this way, they "herd" schools of fish and scoop them up in their huge throat pouches. This pelican's daily food intake of about 2½ lb (1.2 kg) may comprise of either a few large fish or hundreds of small ones. Adult birds develop a crest, a yellow patch on the foreneck, and a pinkish bloom to the plumage during the breeding season. The males display in groups to attract receptive females. Mated pairs display further and the female selects a nest site. Breeding colonies of up to 30,000 pairs may occur, sometimes far from water.

FACT FILE

RANGE
S Europe, Africa, Asia

HABITAT
Lakes and inland seas

SIZE
55–69 in (140–175 cm)

BROWN PELICAN
Pelecanus occidentalis

U nlike other pelicans, this species plunges for fish from heights of 10–33 ft (3–10 m) above the water's surface. It traps fish in its expanded pouch. Brown Pelicans nest in colonies on the ground or in trees. The male occupies a potential nest site from which he displays, leaving the spot only for brief flights until he attracts a mate. Mating takes place on the site and the male then collects branches and other nesting material to present ritually to his partner. The sexes share the incubation of the 2–3 eggs for 30 days. During the first week after hatching, the adults provide the young with well-digested fish, which they regurgitate into the nest. The young fly from the nest after 11–12 weeks. Brown Pelicans have a variety of calls, including grunts and "pops" in the adults and screams and groans in the chicks.

FACT FILE

RANGE
Pacific and Atlantic coasts of
North and South America;
Galapagos Islands

HABITAT
Shallow coastal waters
and islands

SIZE
43–54 in (110–137 cm)

AUSTRALIAN PELICAN
Pelecanus conspicillatus

The Australian Pelican is a highly sociable bird, feeding, roosting, and nesting in large flocks. Adults have a short gray crest and a yellowish or pinkish throat pouch, which becomes scarlet during courtship. Males display communally, with rituals that include swinging the head and clapping the mandibles of the bill together. After they have paired up, a male and female parade to the nest site. They breed on small islands, often making their nests in bushes, and may lay their eggs at any time of the year. Two weeks after hatching, the young may climb into the adult's pouch to feed on regurgitated food. Pelicans feed predominantly on fish but also catch crustaceans and other aquatic animals.

FACT FILE

RANGE
Australia and Tasmania

HABITAT
Shallow marine and
inland waters

SIZE
59–71 in (150–180 cm)

GREAT CORMORANT
Phalacrocorax carbo

race *lucidus*
"White-breasted Cormorant"

race *sinensis*

race *carbo*

There are several races of the Great Cormorant, showing varying amounts of white in the plumage. They include *P. c. carbo*, which breeds around the North Atlantic, *P. c. sinensis*, which breeds from southern Europe to central Asia, and *P. c. lucidus* of sub-Saharan Africa. In some places, this species is persecuted because of its appetite for fish. Its daily intake of food averages 14–25 oz (400–700 g), equivalent to some 15 percent of its body weight. The birds catch their prey during underwater dives that may last for over a minute. The breeding colonies may number up to 2,000 pairs, although they are usually smaller and often fragmented. They may be located on the coast or inland, on cliffs, slopes, or in trees. The courtship display includes wing-flicking, covering and uncovering the conspicuous white flank patch, and a throwing back of the head.

FACT FILE

RANGE
N Atlantic, Africa, Eurasia, Australasia

HABITAT
Coastal and freshwater

SIZE
31½–39 in (80–100 cm)

ANHINGA
Anhinga anhinga

♂

♀

Like other darters, this bird is much more slender than cormorants, with an elongated neck and a thin, straight bill. It often swims with only the head and neck above the surface, the neck held in an S-shape resembling the curve of a snake. Indeed, Snake Bird is an alternative name for this bird. It hunts for fish in rivers, lakes, and lagoons, spearing them underwater with its sharp bill. Like other darters, it has a hinge mechanism in its neck, enabling it to snap its head forward suddenly to seize prey. Darters have large, broad wings, which are excellent for soaring, and in level flight, they alternate flaps with glides. Their courtship ritual includes an aerial display, in which they plane down to the nesting area from great heights. Gregarious at all times, they breed in groups up to hundreds strong, along with cormorants and herons.

FACT FILE

RANGE
SE U.S.A. to N South America

HABITAT
Inland and brackish waters;
can occur on coasts

SIZE
34 in (86 cm)

Sunbittern
Eurypyga helias
The Sunbittern captures prey from beneath the water with its long straight bill.

Squacco Heron
Ardeola ralloides
This heron can be difficult to spot among marshes.

Marabou Stork
Leptoptilos crumeniferus
The Marabou Stork will scavenge on the kills of other animals and on garbage.

Takahe
Porphyrio mantelli
The Takahe lives in alpine and subalpine tussock grassland.

STALKING AROUND THE MARGINS

The water's edge offers feeding opportunities with a range of potential prey for many birds.

Herons, egrets, storks, ibises, and cranes wade on long legs and catch food in their bills. Herons eat fish and frogs. Spoonbills sweep their broad-tipped bills through shallow water to find a fish or amphibian before snapping it up. Smaller egrets chase small fish fry in the shallows while more heronlike larger species wait for fish to come close. Ibises probe soft mud for worms and insects with their long, curved bills. The huge Marabou Stork feeds on carrion and fish.

Flamingos' angled bills encase a fine mesh filter. Flamingos move their beaks "upside down" in shallow water and sieve out tiny crustaceans.

The long toes of crakes and rails, including moorhens and coots and the massive Takahe, help spread their weight so that they can walk over floating vegetation or soft mud without sinking. They eat tiny fish fry, newts, and a range of waterside invertebrates.

Whooping Crane
Grus americana
This crane feeds on plant and animal food from dry land and shallow water.

Purple Swamphen
Porphyrio porphyrio
This bird climbs marshes to catch the early morning sun.

AMERICAN BITTERN
Botaurus lentiginosus

The American Bittern is a stocky brown heron with a black and brown wing pattern in flight. Juvenile birds have more streaking on the back and breast, but closely resemble adults. Like other bitterns, these birds often point their bills to the sky when threatened and even sway back and forth like reeds in the wind. The male American Bittern's 3-syllable pumping or booming call is probably used to advertise territory as well as to attract females. The nest is a platform woven out of plant material on a mound of vegetation or on the ground among cattails or other emergent plants. The population of this species is declining all over its range because of marsh drainage and other habitat changes.

FACT FILE

RANGE
C Canada to C U.S.A.;
winters S U.S.A., Caribbean,
Mexico, Central America

HABITAT
Fresh and saltwater marshes,
swamps and bogs

SIZE
25 in (64 cm)

BLACK-CROWNED NIGHT HERON

Nycticorax nycticorax

juv

The Black-crowned Night Heron is an extremely widespread species. It lives up to its name in being active mainly in the evening and through the night, hunting for frogs and toads, small fish, and crustaceans. Breeding takes place in colonies in trees, with up to 30 nests crowded into a single tree. Though Black-crowned Night Herons sometimes nest with other heron species, they are usually segregated in their own part of the colony. Each nest is an untidy platform of twigs, initiated by the male and completed by the female while the male continues to supply the twigs. When feeding their young, the birds hunt more frequently during the daytime.

FACT FILE

RANGE
Worldwide except N temperate regions and Australia

HABITAT
Salt, brackish, and freshwater wetlands

SIZE
23–25½ in (58–65 cm)

GREEN-BACKED HERON
Butorides striatus

race *rogersi*

race *virescens*

race *sundevalli*

FACT FILE

RANGE
Worldwide tropics and
subtropics

HABITAT
Freshwater and
coastal wetlands

SIZE
16–19 in (40–48 cm)

The Green-backed, or Striated, Heron shows great differences in plumage across its vast geographical range, with races that include *B. s. sundevalli* of the Galapagos Islands, *B. s. rogersi* of northwest Australia, and the American race *B. s. virescens*. It is usually a secretive species, feeding mainly at night and creeping among dense marsh vegetation. However, it sometimes feeds by day at urban ponds, stalking around the edges or perching on jetties and boats. Colonial breeding is rare and restricted to small, loose groups. Most nests are solitary and placed in low bushes or trees, often overhanging water. The 2–4 eggs may be seen through the flimsy nest of twigs.

SQUACCO HERON
Ardeola ralloides

juv

As a result of its overall buff plumage and retiring
habits, the Squacco Heron is a difficult species to
spot among marsh vegetation. As it takes flight, however,
there is a startling change in appearance, with the white
wings, rump, and tail becoming clearly visible. Like many
herons and egrets, the Squacco Heron gains an elegant
crest and ornamental back plumes for the breeding season.
In addition, its bill and legs, which are yellowish-green in
winter, change to blue and pink respectively. Breeding takes
place in mixed colonies with related species, the nests being
built in low trees, bushes, or reeds. Each clutch contains
3–6 eggs laid on a platform of twigs or reeds.

FACT FILE

RANGE
S Europe, SW Asia, Africa

HABITAT
Freshwater marshes,
swamps, and lakes

SIZE
17–18½ in (44–47 cm)

CATTLE EGRET
Bubulcus ibis

winter

FACT FILE

RANGE
S Eurasia, Africa,
Australasia, S U.S.A. and
northern S America

HABITAT
Freshwater wetlands,
farmland and open country

SIZE
19–21 in (48–53 cm)

In the 20th century, the Cattle Egret colonized North and South America, Australia, and New Zealand. It commonly feeds around grazing cattle and wild game herds, which flush prey, such as grasshoppers, beetles, and lizards, out from cover as they move. It also feeds on many invertebrates that are parasites of cattle, such as ticks. During the breeding season, Cattle Egrets sport long, buff feathers on their heads, backs, and chests, and their bills and legs become pinkish-red. They nest in dense colonies, often in company with other species. The nests can be virtually touching, with as many as 100 in a single tree.

LITTLE EGRET
Egretta garzetta

winter

A century ago, this species was nearly exterminated from parts of its range due to the hat trade, which was eager for the 10 in (25 cm) long plumes that form part of the bird's breeding plumage. These feathers have a silky appearance and are soft to touch. However, changing fashions and active campaigning by bodies, such as the Audubon Society and RSPB, have removed this threat. Little Egret colonies can contain many hundreds or even thousands of pairs. They are remarkably agile when climbing over twigs and build their nests in trees. The young leave the nest before fledging to perch on nearby branches.

FACT FILE

RANGE
S Eurasia, Africa, and Australasia

HABITAT
Shallow fresh and brackish water, also estuaries and coasts

SIZE
21½–25½ in (55–65 cm)

PURPLE HERON
Ardea purpurea

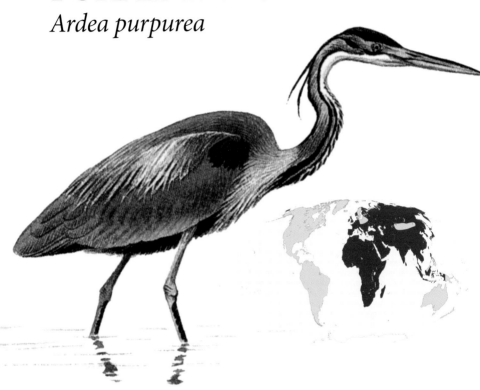

FACT FILE

RANGE
S Eurasia, Africa

HABITAT
Freshwater wetlands
with extensive emergent
vegetation

SIZE
31–35½ in (78–90 cm)

The Purple Heron has longer toes than other, similar-sized species. These enable the bird to walk over floating vegetation without sinking and to stride over thick bushes without having to grasp individual twigs. The large feet are clearly visible in flight and form a useful clue to identification. Most Purple Herons nest in small colonies of up to 20 pairs in dense stands of reeds or rushes growing in shallow water. Occasionally, they choose low trees. The nest is a platform of reeds or twigs and extra platforms are often constructed nearby. These are used by the non-incubating bird and, later, by the developing young.

GRAY HERON
Ardea cinerea

juv

This species haunts a variety of waterside habitats where it catches fish, frogs, and small mammals in its long daggerlike bill. It breeds farther north than any other heron and some populations suffer high rates of mortality in severe weather. However, the birds have a marked capacity for recovering their numbers after such losses. Gray herons usually build their nests in tall trees up to 80 ft (25 m) above the ground. Colony size is variable, with about 200 being the maximum in most areas, although concentrations of more than 1,000 breeding birds have been recorded, A pair readily lays another clutch of eggs if their first is destroyed; they can repeat this 2–3 times.

FACT FILE

RANGE
Widespread in Eurasia
and Africa

HABITAT
Shallow freshwaters of
all types; also coasts,
especially in winter

SIZE
35½–38½ in (90–98 cm)

GREAT BLUE HERON
Ardea herodias

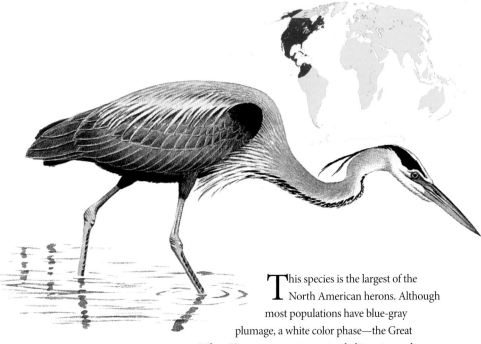

This species is the largest of the North American herons. Although most populations have blue-gray plumage, a white color phase—the Great White Heron—occurs in marine habitats in southern Florida and Cuba. Great Blue Herons forage largely in water, walking slowly or standing and waiting for prey to draw near. Occasionally, they hunt more actively and run, hop, and flick their wings. They feed mostly on fish, aquatic invertebrates, frogs, and small mammals. Breeding adults have plumes on their backs and perform elaborate courtship displays, which include feather-fluffing, stereotyped postures, twig-shaking, and display flights. They nest in colonies, usually in remote areas, with nests in trees up to 130 ft (40 m) above the ground.

FACT FILE

RANGE
S Canada to Central
America, Caribbean,
Galapagos Islands; winters S
U.S.A. to N South America

HABITAT
River edges, marshes,
swamps, mudflats

SIZE
40–50 in (102–127 cm)

GOLIATH HERON
Ardea goliath

The largest of the heron family, this huge bird can wade through much deeper water than related species. Though it eats some amphibians, it feeds mainly on fish, including specimens up to 4½–6½ lb (2–3 kg) in weight. It typically occurs alone or as a single pair. The Goliath Heron nests on the ground in reeds or on a low bush standing in the water. The nest itself is built from reed stems or twigs and is up to 3 ft (1 m) across. The 2–3 pale blue eggs are incubated by both parents for about 4 weeks and the young take a further 6 weeks to fledge.

FACT FILE

RANGE
S and E Africa,
Madagascar, S Iraq

HABITAT
Coastal and inland wetlands
with extensive shallows

SIZE
53–59 in (135–150 cm)

WHITE STORK
Ciconia ciconia

FACT FILE

RANGE
Temperate and S Europe, N
Africa, S and E Asia; winters
in Africa, India, and S Asia

HABITAT
Open, moist lowlands and
wetlands, generally close to
human habitation

SIZE
39–45 in (100–115 cm)

The natural nest sites of the White Stork are trees and cliff ledges, but a great many pairs, certainly the majority in Europe, nest on buildings. Despite much human affection for the species, its population has declined in many areas because of the drainage of marshy feeding areas and the use of pesticides on farmland where many also feed. The main display of the White Stork involves bill-clappering, in which the mandibles are clapped rapidly together. This is a form of greeting between a pair of White Storks throughout the nesting period.

MARABOU STORK
Leptoptilos crumeniferus

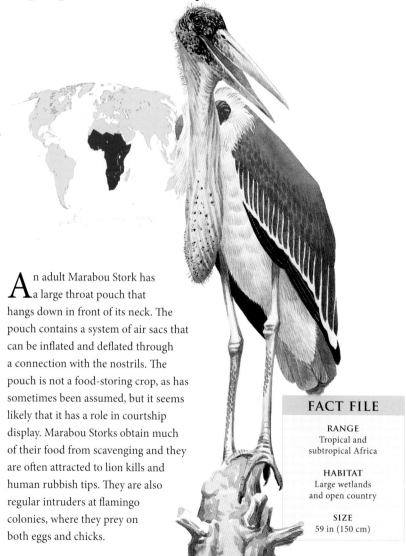

An adult Marabou Stork has a large throat pouch that hangs down in front of its neck. The pouch contains a system of air sacs that can be inflated and deflated through a connection with the nostrils. The pouch is not a food-storing crop, as has sometimes been assumed, but it seems likely that it has a role in courtship display. Marabou Storks obtain much of their food from scavenging and they are often attracted to lion kills and human rubbish tips. They are also regular intruders at flamingo colonies, where they prey on both eggs and chicks.

FACT FILE

RANGE
Tropical and subtropical Africa

HABITAT
Large wetlands and open country

SIZE
59 in (150 cm)

WOOD STORK
Mycteria americana

Wood Storks are gregarious birds and sometimes flocks of 100 or more soar high above the ground on thermals, flying with their heads and necks extended. Wood Storks feed mostly on fish, but also take reptiles, frogs, and aquatic invertebrates. They forage in water up to their bellies, moving their open bill from side to side. When a Wood Stork encounters prey, the bird snaps its bill shut. The birds nest colonially, with up to 25 nests per tree, placed up to 100 ft (30 m) above the ground.

FACT FILE

RANGE
SE U.S.A., Mexico, Central America, W South America to N Argentina

HABITAT
Ponds, marshes, swamps, lagoons

SIZE
34–40 in (86–102 cm)

Scarlet Ibis
Eudocimus ruber

juv

This gregarious species forages and roosts communally, often with several heron and egret species. A sunset flight of Scarlet Ibises to their roost in a mangrove thicket makes a magnificent sight. Their range overlaps in several places with the White Ibis *E. albus* and hybridization occurs. Scarlet Ibises feed mainly on crabs, mollusks, and other invertebrates—they probe for them on mudflats with their curved bills. They also feed on fish, frogs, and insects. They breed colonially, constructing nests of twigs and sticks.

FACT FILE

RANGE
Venezuela, Colombia, coastal Guianas and Brazil; Trinidad

HABITAT
Coastal swamps, mangroves, lagoons, estuaries, mudflats

SIZE
24 in (61 cm)

GLOSSY IBIS
Plegadis falcinellus

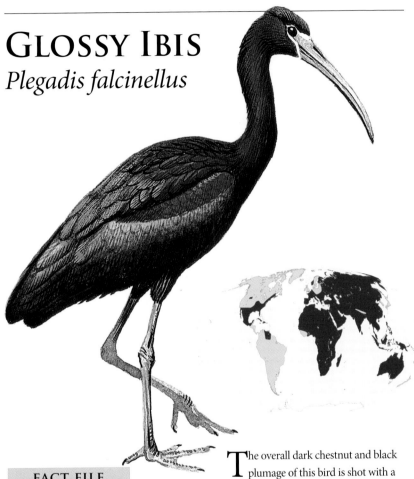

FACT FILE

RANGE
Widespread but
scattered in Central
America, Africa,
S Eurasia, and Australasia

HABITAT
Shallow freshwater and
coastal wetlands

SIZE
22–26 in (56–66 cm)

The overall dark chestnut and black plumage of this bird is shot with a purple and green iridescence that gives the species its name. Glossy Ibises breed in colonies, sometimes of thousands of pairs, almost always in association with herons and egrets. However, they are clearly less tolerant of disturbance than these other species and have declined in many areas. Outside of the breeding season, flocks of Glossy Ibises feed in shallow wetlands and open fields. They roost communally, often sharing waterside trees with herons.

WHITE SPOONBILL
Platalea leucorodia

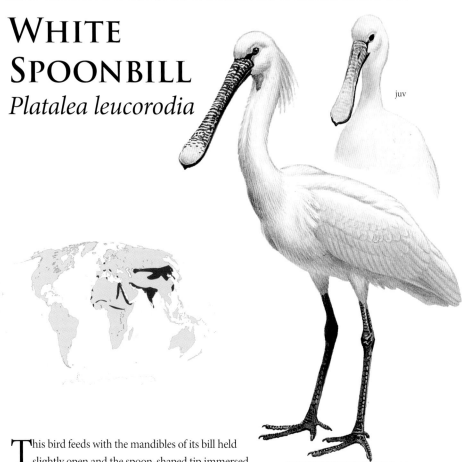

juv

This bird feeds with the mandibles of its bill held slightly open and the spoon-shaped tip immersed in shallow water. As the bird swings its head from side to side, the bill makes scything movements below the surface in search of shrimp and other aquatic life. Feeding usually takes place with small groups wading in lines, which may increase efficiency if one bird is able to catch the prey disturbed by the next bird. White Spoonbills are sensitive to disturbance, and this, as well as the drainage and pollution of their marshland habitats, has caused the population to decline in many areas.

FACT FILE

RANGE
Temperate and S Eurasia, India, tropical W and NE Africa

HABITAT
Shallow fresh, brackish, and saltwater wetlands

SIZE
31–35 in (79–89 cm)

ROSEATE SPOONBILL
Ajaia ajaja

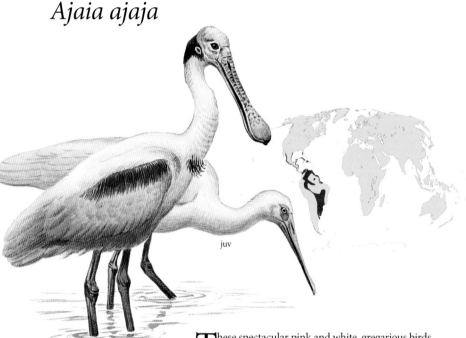

juv

These spectacular pink and white, gregarious birds, are often seen foraging or flying in small flocks. They sweep their bills through shallow water to catch small fish, crustaceans, and other aquatic invertebrates, and also eat some plant material. In the U.S.A., they were persecuted by plume-hunters for the millinery trade. Although they suffer locally from habitat destruction, their numbers and range in south-central U.S.A. have recently increased. Roseate Spoonbills nest in small colonies, often with a variety of heron species. They are monogamous and have elaborate courtship behavior that includes the presentation of twigs, flight displays, and bill-clappering. Their nests of sticks and twigs are constructed in bushes, trees, or reeds or, occasionally, on the ground.

FACT FILE

RANGE
Central America, Colombia, Ecuador, E Peru, Bolivia, N Argentina

HABITAT
Marshes, lagoons, mangroves, mudflats

SIZE
32 in (81 cm)

GREATER FLAMINGO
Phoenicopterus ruber

race *roseus*

The extraordinary slender build and the curious bill of a flamingo are both adaptations for feeding. Small shrimp and other crustaceans are sifted from the bottom mud by a pumping action of the tongue, which forces mud and water through the comblike lamellae on either side of the bill. The head is held upside down at the end of the very long neck and swung to and fro as the bird walks slowly forward. The great length of the legs and the neck allows the bird to feed in much deeper water than other species, so avoiding competition. There are 2 races of the Greater Flamingo. A redder race, *P. r. ruber*, with a bill of orange-pink and black with a yellowish base, which breeds in the Caribbean, and the second race *P. r. roseus*, with paler plumage and a pink and black bill, which breeds in the Old World.

FACT FILE

RANGE
S Europe, SW Asia, E, W, and N Africa, West Indies, Central America, Galapagos Islands

HABITAT
Saline or alkaline lakes, lagoons, and deltas

SIZE
49–57 in (125–145 cm)

LIMPKIN

Aramus guarauna

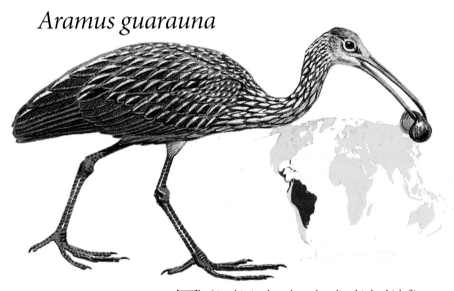

FACT FILE

RANGE
SE U.S.A., Antilles, S Mexico
E of Andes to N Argentina

HABITAT
Marshes, open or wooded
swamps, mangroves

SIZE
23–28 in (58–71 cm)

The Limpkin is a long-legged wading bird, which flies with its neck and head extended. It is largely terrestrial, has rounded wings, and makes only short, low flights. The call, often given at night, consists of a variety of shrieks and screams, or melancholy wails and cries. Limpkins walk with an undulating tread so that they appear to be limping—hence their common name. Limpkins probe and grasp with their laterally compressed bill, which is well adapted for removing freshwater snails from their shells. The birds also feed on mussels, and occasionally on seeds, small reptiles and frogs, insects, worms, and crayfish. Although usually solitary, they sometimes hunt in small groups. The nest is a woven mat of reeds and sticks built in or under shrubs and trees near water. The normal clutch contains 6–7 brown-spotted buff eggs, which are incubated for about 20 days. The chicks leave the nest on the day that they hatch but remain with their parents.

BLACK CROWNED CRANE
Balearica pavonina

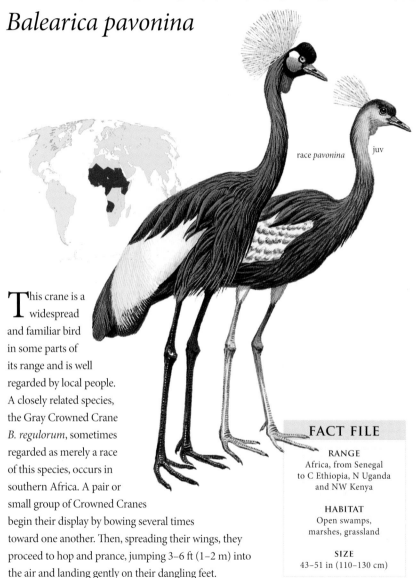

race *pavonina*   juv

This crane is a widespread and familiar bird in some parts of its range and is well regarded by local people. A closely related species, the Gray Crowned Crane *B. regulorum*, sometimes regarded as merely a race of this species, occurs in southern Africa. A pair or small group of Crowned Cranes begin their display by bowing several times toward one another. Then, spreading their wings, they proceed to hop and prance, jumping 3–6 ft (1–2 m) into the air and landing gently on their dangling feet.

FACT FILE

RANGE
Africa, from Senegal to C Ethiopia, N Uganda and NW Kenya

HABITAT
Open swamps, marshes, grassland

SIZE
43–51 in (110–130 cm)

SANDHILL CRANE
Grus canadensis

race *tabida*
"Greater Sandhill Crane"

rust-stained

juv

Sandhill Cranes may congregate at migration stopovers or on the wintering grounds in flocks of hundreds, creating an extraordinary din with their grating, rattling, trumpetlike calls. They are ground-gleaners, foraging for a wide variety of plants, aquatic invertebrates, insects, worms, frogs, and small mammals. There are 5 races, varying in size. The race illustrated, the Greater Sandhill Crane *G. c. tabida*, breeds from southwest British Columbia to northern California or Nevada. The rusty-brown color of adults in some areas is due to their habit of probing with their bills into mud and soil containing reddish iron oxide. When they preen, they transfer the stain to their plumage.

FACT FILE

RANGE
Breeds NE Siberia,
Alaska, Canada, N U.S.A.;
winters S U.S.A. to C
Mexico; small resident
populations in Florida and
Mississippi, Cuba

HABITAT
Tundra, marshes,
grassland, fields

SIZE
42 in (107 cm)

BROLGA
Grus rubicundus

juv

The Brolga is a large, gray crane native to damp habitats in parts of Australia. It has declined in abundance in the southeastern section of its range because of the drainage of shallow wetlands. It feeds mainly on tubers, which it digs up from the mud with its bill, but will also eat grain, insects, mollusks, and small vertebrates. Brolgas perform elaborate dancing displays, bowing and stretching, advancing and retiring, with much wing-flapping and trumpeting. They also leap high and plane down again, and hurl pieces of grass or twigs. Elements of their display have been incorporated into Aboriginal dances.

FACT FILE

RANGE
N and E Australia
S to Victoria

HABITAT
Shallow, open swampland
and wet grassland

SIZE
37½–49 in (95–125 cm)

WHOOPING CRANE
Grus americana

juv

In flight, an adult Whooping Crane shows black outer flight feathers that contrast with its otherwise white plumage. Like all cranes, it flies with its head and neck extended. With a total population of fewer than 200 individuals, it is one of the world's most endangered birds. Attempts are being made to start new populations by using Sandhill Cranes as foster parents. Whooping Cranes are opportunist feeders, taking a wide variety of plant and animal food from dry land and shallow water. The Whooping Crane's courtship ceremony involves elaborate dances, and its most common call is a whooping or trumpeting *ker-loo*.

FACT FILE

RANGE
Breeds C Canada; winters coastal Texas

HABITAT
Breeds on muskeg and prairie pools; prairie and stubble fields on migration; winters on coastal marshes

SIZE
51 in (130 cm)

COMMON CRANE
Grus grus

chick

juv

Like the other members of the crane family, the Common Crane is famous for its dancing display. This is performed not only during courtship but also by small groups of birds. The birds walk around in circles, bowing, bobbing, pirouetting, and stopping, bending to pick up small objects and toss them over their heads. All these rituals are interspersed with graceful jumps high into the air. The species' main call is a loud trumpeting, which is uttered by courting pairs as they stand facing each other with outspread wings, necks stretched, and bills pointing skyward. This trumpeting is also heard from the V-formations of migrating flocks.

FACT FILE

RANGE
N temperate Eurasia; winters S to N Africa, India, and Southeast Asia

HABITAT
Breeds in open forest swamps and moorland bogs; winters in shallow wetlands

SIZE
43–47 in (110–120 cm)

WATER RAIL
Rallus aquaticus

FACT FILE

RANGE
Eurasia, from Iceland to
Japan; N Africa

HABITAT
Marshes, swamps,
wet meadows

SIZE
11 in (28 cm)

With its very narrow body, the Water Rail can slip noiselessly through reeds in its habitat. Though one of the most widespread and common rails, it is a shy, skulking bird, particularly in summer, but when the marshes freeze in winter, it often emerges to forage across open ground. The Water Rail will tackle a wide variety of prey, from insects to amphibians and nestling birds, and may leap up to 3 ft (1 m) to snatch a dragonfly from its perch above the water. Like almost all rails, it can swim well when necessary. Though rarely seen during the breeding season, it is often heard then and at other times as it utters a wide range of loud, blood-curdling groans and piglike grunts and squeals in a process known as "sharming." The nest is well hidden, often in a tussock of rushes, and the 5 or more highly mobile chicks, covered in black down, are soon able to race around the marsh after their parents.

CLAPPER RAIL
Rallus longirostris

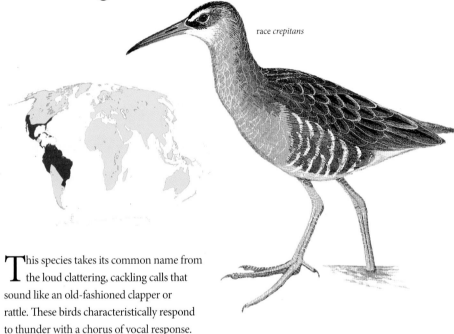

race *crepitans*

This species takes its common name from the loud clattering, cackling calls that sound like an old-fashioned clapper or rattle. These birds characteristically respond to thunder with a chorus of vocal response. Once a common sight on salt marshes throughout much of America, the Clapper Rail has suffered badly from habitat loss and hunting over the last 50 years. Even without these pressures, it leads a precarious existence, nesting among salt marsh grasses where it is liable to lose whole clutches of eggs to a single high spring tide. When the eggs do survive to hatch, however, the downy, black chicks, like the adults, can swim well and even submerge by grasping underwater plants with their feet when threatened. Clapper Rails feed mainly on crabs and smaller crustaceans, supplemented by seeds and tubers. In severe winters, the northern birds move south but most are resident.

FACT FILE

RANGE
S Canada to N Peru;
Caribbean islands

HABITAT
Brackish estuaries and
salt marsh, especially with
glasswort plants

SIZE
12½–18 in (32–46 cm)

WEKA
Gallirallus australis

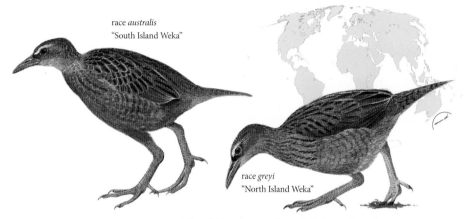

race *australis*
"South Island Weka"

race *greyi*
"North Island Weka"

Wekas are very large flightless rails with powerful bills and feet. They are inquisitive and omnivorous birds, with a diet that includes grass, seeds, fruit, mice, birds, eggs, beetles, and snails. They frequently raid garbage bins and even enter fowl runs to eat the chicks. They can be voracious predators of other ground-dwelling birds and have upset the natural population balance on some of the islands to which they have been introduced. Conversely, Wekas kill many rats. The North Island race *G. a. greyi* is restricted to the Gisborne district, except for some reintroductions elsewhere. The paler Buff Weka *G. a. hectori* became extinct in the South Island in the 1920s but survives in Chatham Island where it was introduced in 1905. The race *G. a. australis* is quite common on South Island and a further race *G. a. scotti*, survives on Stewart Island.

FACT FILE

RANGE
New Zealand

HABITAT
Scrubland, forest edge

SIZE
21 in (53 cm)

SORA
Porzana carolina

juv

Many thousands of these crakes were victims to American hunters each year as they migrated to their coastal wintering grounds; now the birds face a threat from the destruction of their breeding marshes. Though they are still among the most abundant and widespread of American rails, there are no longer the dense flocks of Soras huddled on salt marshes, islands, and even seagoing vessels to be seen. In its summer haunts, the Sora is a shy bird, rarely seen and usually detected by its call, which is a loud, staccato *cuck* accompanied by sneezing and whinnying sounds. It lays a large clutch, usually of 8–12 eggs, which is incubated by both sexes.

FACT FILE

RANGE
Canada to Venezuela and Peru, including Caribbean islands

HABITAT
Freshwater marshes, also wet meadows in prairie regions; winters by salt and brackish waters

SIZE
9 in (23 cm)

COMMON MOORHEN
Gallinula chloropus

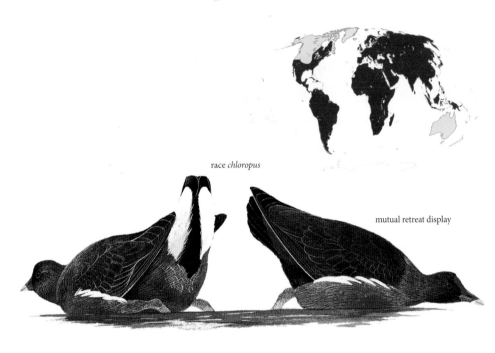

race *chloropus*

mutual retreat display

FACT FILE

RANGE
Throughout temperate and tropical Eurasia, Africa, North and South America; but not Australasia

HABITAT
Small ponds, rivers, wet marshes

SIZE
14 in (35 cm)

Possibly the most abundant species of rail in the world, the Common Moorhen is an unlikely candidate for success—it is a poor flier and is barely an adequate swimmer, preferring to search for food on foot with a delicate, high-stepping gait. Male Common Moorhens are highly territorial and will fight with their large feet; many are badly injured in the process. The chicks become independent very quickly and often help their parents feed a second brood.

PURPLE SWAMPHEN
Porphyrio porphyrio

race *porphyrio*

The Purple Swamphen has earned itself a bad reputation in some areas due to its habit of feeding on the tender shoots of cultivated rice. Largely vegetarian, it may become extremely numerous where food is plentiful; scores of birds can often be seen in the early morning when they climb the marsh vegetation to sunbathe as the first rays of dawn penetrate the mist. Usually, a single pair seems to do all the nesting; occasionally, a second female may lay in the nest.

FACT FILE

RANGE
Europe, Africa, S Asia,
Southeast Asia to Australasia

HABITAT
Reed beds, scrubby marsh
margins, tussock swamps

SIZE
18 in (45 cm)

TAKAHE
Porphyrio mantelli

FACT FILE

RANGE
New Zealand, now restricted
to Murchison and Kepler
ranges in SW South Island;
introduced to Mana Island

HABITAT
Mountain tussock
grassland and (rarely)
adjacent woodland

SIZE
25 in (63 cm)

Like many other flightless island birds, this large rail has suffered a catastrophic decline in the face of competition and predation by introduced animals. It was first described scientifically in 1849. Its numbers dwindled rapidly until, by 1900, it was believed to be extinct. In 1948, it was rediscovered living in remote alpine and subalpine tussock grassland in southwest South Island. Here, it feeds on the seed heads and tender bases of the grasses. Today, it has to compete for the grass with introduced deer and generally loses the contest. Takahes mate for life, nesting on the ground among tussocks. Introduced predators, such as stoats, take a heavy toll of the eggs and young, and the breeding success rate is low. Captive breeding, conservation, and predator control have helped to preserve the species, but the Takahe remains endangered.

EURASIAN (BLACK) COOT
Fulica atra

aggressive encounter

The most widespread and common of all the coots, this species is a bold and familiar bird in Europe. The bird feeds mainly on aquatic plants, but it also eats aquatic insects. During the spring breeding season, both sexes are extremely territorial and the slightest violation of a boundary by another coot will provoke a tremendous show of strength as the intruder is attacked with both claws and bill. In winter, coots become very sociable, flocking together in hundreds or even thousands on food-rich lakes and rivers. Juvenile birds are much paler than the adults, with lighter, duller legs, a grayish bill, and brown eyes—they also they lack the adults' white frontal shield.

FACT FILE

RANGE
Widespread through Europe, Asia, Japan, and Australasia

HABITAT
Large ponds, lakes, and marshes, usually at low altitude

SIZE
17 in (43 cm)

Sunbittern
Eurypyga helias

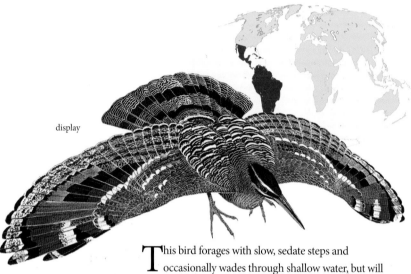

display

This bird forages with slow, sedate steps and
occasionally wades through shallow water, but will
fly across deeper pools. It feeds mainly on insects, spiders,
small crustaceans, frogs, and small fish, seizing them from
the mud and rocks, from beneath the water or among
damp litter on the forest floor, with heronlike jabs of its
long straight bill. Though the Sunbittern spends most of its
time in the shade, it is sometimes found in sunny clearings
when its spectacular display is seen to the best advantage.
The bird lowers its neck, spreads its wings, and raises and
fans its tail, suddenly revealing the striking areas of orange-
chestnut surrounded by pale orange-buff on its primary
wing feathers. This coloring resembles a sunset—hence the
bird's common name. The Sunbittern lays 2 pale eggs in a
bulky nest of stems and leaves in a tree, 10–20 ft (3–6 m)
above the ground. Both sexes incubate the eggs for about
27 days; the nestling period is about 21 days.

FACT FILE

RANGE
S Mexico to Bolivia
and C Brazil

HABITAT
Forested streams,
rivers, lakes

SIZE
18 in (46 cm)

WATTLED JACANA
Jacana jacana

race *hypomelaena*

race *jacana* juv

race *jacana*

The Wattled Jacana runs across the surface of floating leaves, searching for snails and other invertebrates and pecking pieces of vegetation. It also frequents damp pastures. It flies fast and low with short glides. The race *J. j. jacana* is found over most of the species' South American range and the dark race *J. j. hypomelaena* occurs from western Panama to northern Colombia. Females mate with more than one male in each season and compete vigorously with one another for the opportunity to mate. Since the males incubate and care for the young, one female may even destroy the eggs of a successful rival in order to gain access to a male. Studies in Panama showed that each female's territory overlapped with the territories of up to 3 males.

FACT FILE

RANGE
Panama S to C
Argentina; Trinidad

HABITAT
Tropical freshwater ponds,
marshes, streams with
emergent vegetation

SIZE
8–9 in (20–23 cm)

Musk Duck
Biziura lobata
The male Musk Duck has a distinctive lobe of skin, which hangs down from the lower mandible.

Marbled Teal
Marmaronetta angustirostris
The Marbled Teal's patterned plumage gives it its name.

Hawaiian Goose
Branta sandvicensis
The Hawaiian Goose's feet are only partially webbed.

Hooded Merganser *Mergus cucullatus*
The male Hooded Merganser has a crest on his head that can be raised, making the head seem larger.

Ducks, Swans, and Geese

Black Swan
Cygnus atratus
The Black Swan appears in huge flocks.

Collectively known as "wildfowl," these birds are found worldwide with a large range of species in the northern hemisphere. Many of these birds breed in vast tundra wetlands in the Arctic, in a brief summer of 24-hour daylight, but they migrate south to avoid the extreme winter weather.

Swans are long-necked, long-winged, short-legged birds, with stout, wedge-shaped bills suitable for grazing, but also grasping underwater vegetation and small aquatic creatures. Geese have shorter necks and stouter, shorter bills for plucking grass and sedges and breaking into roots. There are 2 main groups. The "gray" geese are mostly dull gray-brown. The "black" geese are more boldly patterned. Ducks have more variety—some are surface-feeders, while others dive for vegetable or invertebrate food, and others, with serrated "saw bills," catch fish. Still others, live at sea and feed on shellfish and crustaceans.

Egyptian Goose
Alopochen aegyptiacus
This goose uses a variety of nest sites.

BLACK SWAN
Cygnus atratus

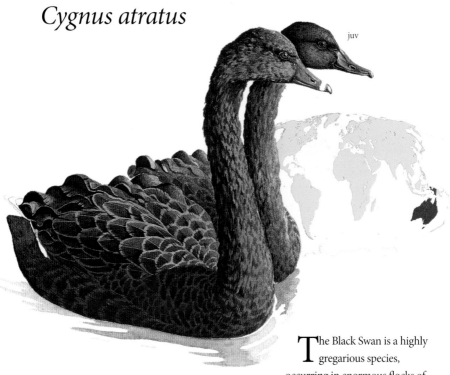

juv

The Black Swan is a highly gregarious species, occurring in enormous flocks of up to 50,000 strong. It breeds in colonies of hundreds of pairs and the nests are placed just beyond pecking distance from one another. Like many Australian waterbirds, the Black Swan can delay breeding in times of drought but can quickly take advantage of sudden rain and flood conditions. Long-distance movements have also been recorded in response to such extremes of climate. The nest is typically a pile of vegetation, up to 3 ft (1 m) across at the base, built in shallow water. The 5–6 pale green eggs are incubated for up to 40 days by both parents. In most other wildfowl, the female carries out the incubation alone.

FACT FILE

RANGE
Australia; introduced into
New Zealand

HABITAT
Large freshwater and
brackish marshes and
lagoons; also estuaries
and coastal bays

SIZE
45–55 in (115–140 cm)

MUTE SWAN
Cygnus olor

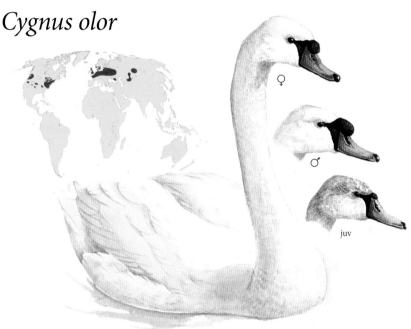

♀

♂

juv

The Mute Swan is a familiar sight in parks in Britain where it has long been protected. Much wilder populations occur in continental Europe and Asia, forming flocks on lakes and estuaries. The species is misnamed, since it utters a variety of snoring, snorting, and hissing calls and its wings produce a loud, throbbing sound in flight. The males (called cobs) have larger knobs on their bills than females (pens), especially in the breeding season. A pair of Mute Swans will vigorously defend a territory around the mound of vegetation that forms their nest. However, the birds breed colonially in a few places and there the strong territorial instinct is reduced so much that their nests are only about 3 ft (1 m) apart.

FACT FILE

RANGE
Temperate Eurasia; introduced to parts of North America, South Africa, and Australia

HABITAT
Lowland freshwater lakes and marshes; coastal lagoons and estuaries

SIZE
49–61 in (125–155 cm)

TUNDRA SWAN
Cygnus columbianus

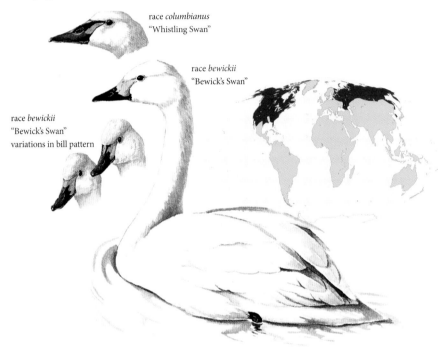

race *columbianus*
"Whistling Swan"

race *bewickii*
"Bewick's Swan"

race *bewickii*
"Bewick's Swan"
variations in bill pattern

FACT FILE

RANGE
Breeds Arctic North America
and U.S.S.R.; winters S to
temperate latitudes

HABITAT
Breeds on marshy tundra;
winters on freshwater
marshes and estuaries

SIZE
47–59 in (120–150 cm)

Three races of the all-white Tundra Swan are recognized. These include the Whistling Swan *C. c. columbianus* of North America and the Bewick's Swan *C. c. bewickii* of Eurasia, both illustrated. The amount of yellow on the bill of the Bewick's Swan is extremely variable and can in fact be used to identify individual birds. Most swans return each winter to the same locality and there is great faithfulness between pairs. The birds build their nest on marshy ground and lay a clutch of 3–5 eggs. The young swans migrate with their parents and they spend their first winter together as a family.

GRAYLAG GOOSE
Anser anser

race *anser*

race *rubrirostris*

The Graylag Goose was once bred widely
throughout western Europe but the drainage of
marshes and deliberate persecution caused a dramatic
reduction in its breeding range. However, it is now
expanding again and some wintering flocks contain tens
of thousands of geese. There are 2 races. *A. a. anser* breeds
in Europe, while the pink-billed *A. a. rubrirostris* breeds in
Asia. They overlap in eastern Europe and western U.S.S.R.
Large winter flocks on farmland may damage crops and
pasture. The bill of the Graylag Goose is adapted for probing
in marshy ground and pulling up roots and stems; it is also
ideal for grazing and slicing pieces off turnips and potatoes.

FACT FILE

RANGE
Mainly temperate latitudes
of Eurasia

HABITAT
Freshwater marshes and
open water; often winters
on farmland

SIZE
29½–35½ in (75–90 cm)

MAGELLAN GOOSE
Chloephaga picta

race *leucoptera*
"Greater Magellan Goose"
♀

race *picta*
"Lesser Magellan Goose"
white phase
♂

race *picta*
barred phase
♂

FACT FILE

RANGE
S South America and the
Falkland Islands

HABITAT
Grassland and marshes

SIZE
23½–25½ in (60–65 cm)

As grazing birds, Magellan Geese have come into conflict with sheep farmers in South America. Bounties have been offered for both birds and eggs, but though large numbers are killed and eggs are systematically removed, the species is not under serious threat. The Magellan Goose's habitats are extensive and the human population is sparse. Though pressure on the birds is severe in some areas, in others, the geese are unmolested. The white and gray male has a whistling call used in display, to which the rufous and gray-brown female responds with a lower cackling. The illustration shows the 2 races of the Magellan Goose: the Lesser *C. p. picta*, with its white and barred phases, and the Greater *C. p. leucoptera*.

MAGPIE GOOSE
Anseranas semipalmata

♀

feeding chick

With its long neck, long legs, and bold pied plumage, the Magpie Goose is one of the most distinctive of all wildfowl. It has only partial webbing between the toes and the male has a pronounced dome to the top of the head. The natural food of the Magpie Goose includes aquatic plants and seeds. In northern Australia, the birds have damaged rice crops, both by grazing and trampling, but careful adjustment of water levels in the rice paddies has helped to reduce the problem. The nest is a large platform of vegetation, plucked off and trampled in shallow water, before being built up into a mound. Uniquely among wildfowl, the parents feed their young bill to bill.

FACT FILE

RANGE
N Australia and
S New Guinea

HABITAT
Overgrown swamps
and lagoons and
adjacent farmland

SIZE
29½–33½ in (75–85 cm)

Snow Goose
Anser caerulescens

race *atlanticus*
"Greater Snow Goose"
snow phase

There are few more magnificent sights than a flock of tens of thousands of Snow Geese. Glistening white, the birds descend like snowflakes, their black wing tips seeming to flicker as they turn before landing. The 2 races, the Greater and the Lesser, are distinguished mainly by size. While the adult Greater Snow Goose is always white, the Lesser Snow Goose shows both the white "snow" phase and a dark gray "blue" phase. Snow Geese nest in colonies of many thousands of pairs on the Arctic tundra. In winter, they journey south, some flying non-stop for 2,000 mls (3,500 km) or more to reach the Gulf of Mexico, where harvested rice fields provide ample food.

FACT FILE

RANGE
Arctic North America;
winters on both seaboards
and S to Gulf of Mexico

HABITAT
Breeds on arctic tundra;
winters on freshwater and
salt marshes, farmland

SIZE
25½–33 in (65–84 cm)

HAWAIIAN GOOSE
Branta sandvicensis

By 1952, extensive hunting and introduced predators had reduced the once numerous population of the Hawaiian Goose. Estimated at 25,000 birds at the end of the 19th century, numbers were reduced to as few as 30 individuals. Since then, captive breeding in both Hawaii and England has enabled well over 1,000 birds to be released back into the wild. By 1976, the total population in the wild was around 750. The Hawaiian Goose has only partial webbing on its feet, reflecting its largely terrestrial habits. In its home on the Hawaiian Islands, there is little standing water. The nest is a simple scrape in the ground, usually in the shelter of a rock or a clump of vegetation.

FACT FILE

RANGE
Confined to the
Hawaiian Islands

HABITAT
Sparsely vegetated
slopes of volcanoes,
shrubland, grassland

SIZE
22–28 in (56–71 cm)

EGYPTIAN GOOSE

Alopochen aegyptiacus

FACT FILE

RANGE
Nile Valley and
sub-Saharan Africa

HABITAT
Tropical and
subtropical wetlands

SIZE
25–29 in (63–73 cm)

A large, upright bird, the Egyptian Goose has 2 color phases, one gray-brown on the upperparts, the other red-brown. The species is remarkable for the great variety of nest sites that it uses. Some birds nest on the ground, like most other geese and sheldgeese, often choosing the shelter of bushes or clumps of grass. Other pairs prefer to nest off the ground, using ledges on cliffs or old buildings. Some select the abandoned nests of other birds, often high in the crowns of trees. A typical clutch consists of 5–8 eggs, which are incubated by the female for about 28 days. The downy young that hatch out on ledges or in trees face a perilous tumble to the ground as they leave the nest. The parents do not assist the chicks, but call to them from below until they pluck up the courage to step into space.

RED-BREASTED GOOSE
Branta ruficollis

A remarkable association
exists between this
scarce, strikingly plumaged
goose and various birds of
prey. Red-breasted Geese
nest on the Arctic tundra,
concentrating in small groups
of up to 5 pairs around the nests of
Peregrines, Rough-legged Buzzards, and, sometimes,
large gulls. In defending their own nests against predators,
these aggressive birds keep intruders, such as gulls and
Arctic Foxes, away from the nests of the surrounding
geese. The geese may site their nests within 33 ft (10 m) of
birds of prey, apparently without risk of attack from their
"protectors." The nest is usually placed on a steep bank
or on the top of a low cliff. The total world population is
probably fewer than 25,000 individuals.

FACT FILE

RANGE
Breeds in the Siberian
Arctic; winters close
to the Black, Caspian,
and Aral seas

HABITAT
Breeds on tundra; winters
on open farmland close to
major wetlands

SIZE
21–21½ in (53–55 cm)

CANADA GOOSE
Branta canadensis

race *minima*

race *canadensis*

FACT FILE

RANGE
Arctic and temperate North
America; introduced into
Europe and New Zealand

HABITAT
Lowland wetlands

SIZE
21½–43 in (55–110 cm)

The Canada Goose was formerly divided into 10–12 races, based on distribution, size, and coloring. The smallest races (such as *B. c. minima* from west Alaska) breed in the Arctic; the larger races (such as *B. c. canadensis* from eastern Canada) breed farther south. The 4 smallest races are now considered a separate species—the Cackling Goose. In recent decades, these large birds have successfully moved into parks and many are tame; in some urban areas, they are seen as a nuisance because they are attracted to lawned areas. The bird's nest is usually placed on the ground, often sheltered by vegetation, but tree and even cliff-ledge nest sites have been reported and rafts and other artificial platforms are readily adopted.

WOOD DUCK
Aix sponsa

The remarkable coloration of the male Wood Duck greatly contrasts with the mottled gray-brown of the female. It was the drake's superb plumage, which nearly led to the species' extinction. In the 19th century, its feathers were in great demand for ornament and for fishing flies. However, because protection measures were taken in time, and since the Wood Duck's habit of nesting in tree holes meant that it took readily to nest-boxes, it was possible to restore the species to many areas of eastern North America in which tree clearance and drainage, as well as shooting, had led to its serious decline.

FACT FILE

RANGE
North America: E population breeds S Canada to Florida, W population breeds British Colombia to California; winters in S of breeding range

HABITAT
Freshwater in wooded areas; more open flooded areas in winter

SIZE
17–20 in (43–51 cm)

MUSCOVY DUCK
Cairina moschata

The Muscovy Duck has been widely introduced outside of its native lands as a domestic fowl. Unlike the glossy greenish-black wild birds, the domestic birds may be green-black, gray or white, or a mixture of colors. The knob at the base of a domestic male's bill is red and surrounded by bare red skin. Muscovy Ducks are omnivorous and feed on the leaves and seeds of marsh plants, as well as small reptiles, fish, crabs, and insects. In the wild, these shy birds roost in small groups in trees, safe from ground predators. The only time they leave the shelter of the forest is when the savannas are flooded. Both wild and domestic Muscovy Ducks prefer to nest in holes in trees, and they take readily to nest-boxes.

FACT FILE

RANGE
Central and South America

HABITAT
Freshwater marshes and pools in wooded areas; brackish coastal lagoons

SIZE
26–33 in (66–84 cm)

GREEN-WINGED TEAL
Anas crecca

♀
race *crecca*

♂
race *carolinensis*

♂
race *crecca*

There are 2 closely similar species. The European *A. crecca* has a horizontal white flank stripe and the North American *A. carolinensis* has a vertical white chest stripe. These are male characteristics. Females are indistinguishable. Green-winged Teals can launch themselves almost vertically into the air when disturbed, "springing" off the water—from this comes an old collective noun for a group of these agile and beautiful little ducks—a "spring" of teal. Once in flight, dense flocks will wheel and turn in perfect unison. The male has a characteristic, far-carrying, liquid *preep preep* call. The diet consists of small invertebrates and some seeds—most of which, the birds find by dabbling in the shallows.

FACT FILE

RANGE
Breeds North America,
N Eurasia; winters S U.S.A.,
Central America, Caribbean,
temperate Eurasia,
tropical Asia

HABITAT
Freshwater pools, marshes;
also on estuaries, coastal
lagoons in winter

SIZE
13–15 in (34–38 cm)

COMB DUCK
Sarkidiornis melanotos

race *melanotos*
♀

♂

FACT FILE

RANGE
Sub-Saharan Africa, India,
Southeast Asia, and tropical
South America

HABITAT
Rivers, swamps, lakes

SIZE
22–30 in (56–76 cm)

This bird takes its name from the fleshy blackish knob at the base of the male's bill. This knob has an edge that is similar to the edge of a comb. Comb Ducks are highly mobile, with flocks of 50 or more birds moving to take advantage of flooded conditions in different areas. Outside of the breeding season, flocks of Comb Ducks often consist entirely of males or of females. During the breeding season, each male may have 2 mates and several nesting "pairs" can form small breeding colonies. The nest is usually in a hole in a tree, in an old wall, or on the ground, well concealed in tall vegetation.

TORRENT DUCK
Merganetta armata

race *armata*
"Chilean Torrent Duck"
♂

♀
race *armata*

The Torrent Duck is perfectly adapted to life in fast-flowing water. It has a streamlined shape, powerful legs, a broad, stiff tail used for steering, and sharp claws to enable it to cling to slippery rocks. It feeds by making frequent underwater forays from its perch on a boulder. It forages for caddis fly and other larvae on the river bottom, catches small fish, and picks various food items from the surface. Though few nests have been found, they are always near water in a crevice or hollow of some kind, between rocks or under overhangs. The ducklings are able to tackle the most turbulent water as soon as they hatch. Several races have been described, separated geographically along the 5,000-ml (8,000-km) length of the Andes. The Chilean Torrent Duck *M. a. armata* of central Chile and western Argentina is illustrated.

FACT FILE

RANGE
Andes

HABITAT
Fast-flowing streams

SIZE
17–18 in (43–46 cm)

MALLARD

Anas platyrhynchos

FACT FILE

RANGE
N hemisphere, N of the tropics; introduced to Australia and New Zealand

HABITAT
Wide range of fresh and coastal waters

SIZE
20–25½ in (50–65 cm)

The success of the Mallard, ancestor of most domestic ducks, reflects its supreme adaptability. It can become completely tame in urban areas, relying on human handouts for food, though it is as wild as any wildfowl in other habitats. Mallards feed by dabbling in shallows, or upending to reach greater depths. They are omnivorous, eating both invertebrates and plant matter. Natural nest sites are in thick vegetation close to water, but in urban areas, the birds use holes in trees and buildings and even window ledges.

NORTHERN SHOVELER
Anas clypeata

The bird's distinctive bill acts as a superb filter for food. As the duck swims along with the front half of the bill submerged, it creates a stream of water that flows in at the bill's tip. Rapid tongue action aids the flow, directing the water out at each side with any particles being strained from the water by intermeshing projections on the upper and lower mandibles. Groups of Northern Shovelers often circle slowly as they feed. Their combined paddling action brings food items—tiny seeds and floating animals—to the surface in deep water. Northern Shovelers will also dabble in shallows, selecting larger food items. Here, however, they come into competition with other ducks, which is something that their sieving method normally avoids.

FACT FILE

RANGE
Eurasia and North America S of the Arctic Circle; winters S to the subtropics

HABITAT
Freshwater pools and marshes; also on estuaries and coastal lagoons in winter

SIZE
17–20 in (44–52 cm)

AMERICAN WIGEON
Anas americana

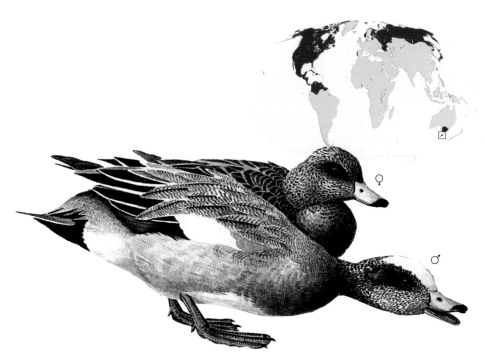

FACT FILE

RANGE
Breeds N and C North America; winters S to Gulf of Mexico

HABITAT
Freshwater marshes, ponds; winters on marshes, coastal lagoons

SIZE
18–22 in (45–56 cm)

American Wigeons are highly gregarious in winter, occurring in flocks of many tens of thousands. In summer, the breeding pairs disperse, spreading throughout the pothole country of the North American prairies and northward into the wooded muskeg of Canada and Alaska. Though classed as a dabbling duck, the American Wigeon obtains only some of its plant food by dabbling in shallow water. More often, tightly packed flocks of birds graze on marshes and pastures. The nest is a shallow cup of grasses in concealing vegetation near water.

AMERICAN BLACK DUCK
Anas rubripes

♂

American Black Ducks feed on plants and invertebrates and nest in dense vegetation close to water. As with some other close relatives of the Mallard, but unlike nearly all other northern hemisphere dabbling ducks, the male lacks any bright nuptial plumage. The species may now be under threat because female American Black Ducks are more attracted by brightly colored male Mallards than by their own kind. Since the recent expansion of the Mallard's range in eastern North America, hybridization has become more common and widespread. Habitat degradation and overhunting pose further threats to the bird's population.

FACT FILE

RANGE
NE North America; winters
S to Gulf of Mexico

HABITAT
Freshwater marshes in
woods; winters on estuaries,
coastal marshes

SIZE
21–24 in (53–61 cm)

NORTHERN PINTAIL
Anas acuta

FACT FILE

RANGE
Eurasia and North America;
winters S to Panama, C
Africa, India, Philippines

HABITAT
Open marshes; winters on
estuaries and coastal lagoons

SIZE
Male: 25–29 in (63–74 cm)
including 4 in (10 cm) central
tail feathers;
Female: 17–25 in
(43–63 cm)

The handsome drake Northern Pintail has long tail streamers, which he cocks out of the water as he swims. This species is strongly attracted to fall stubble fields and a single site may hold more than 500,000 birds. In some parts of North America, wheat and corn is grown specially for them—it is left unharvested in the fall to attract the vast hordes and keep them from damaging more vulnerable crops. Apart from pecking at grain, Northern Pintails feed mainly in the water. They upend, using their long necks to reach deeper than other ducks sharing their range. They scour the bottom of pools and marshes for plant roots and leaves, as well as aquatic invertebrates.

Northern Shelduck
Tadorna tadorna

These large, gooselike birds feed with a scything action of the bill through soft mud and sand, picking up mollusks, crustaceans, and other invertebrates. In early spring, they gather in groups in which the males court the females with melodious whistling calls, rearing up and throwing back their heads. Arguments with neighbors often develop into furious aerial chases. Most Northern Shelducks nest in holes, either underground or in hollow trees. They will also nest under buildings and in straw stacks. The young of several broods often combine into large "crèches," under the care of just a few of the parents. The rest of the adults depart for their molting grounds. Much of the population of northwestern Europe, numbering some 100,000 birds, migrates to molt in the Waddenzee on the north German coast.

FACT FILE

RANGE
Temperate Eurasia; winters S to N Africa, India, China, and Japan

HABITAT
Estuaries, shallow seas

SIZE
23–26 in (58–67 cm)

MAGELLANIC FLIGHTLESS STEAMER DUCK

Tachyeres pteneres

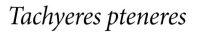

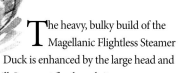

The heavy, bulky build of the Magellanic Flightless Steamer Duck is enhanced by the large head and strong bill. It cannot fly, though it can escape predators by furiously thrashing and paddling over the water with its stubby wings. It is this action, once described as "steaming," that earned the duck its name. The males, in particular, also use their wings to strike one another in combat. The blows are reinforced by large bony knobs on the carpal joints at the bends of the wings. Magellanic Flightless Steamer Ducks feed exclusively on marine invertebrates, especially mussels and crabs. They dive for their food among forests of kelp fronds and also dabble in gravel under shallow water. The nest is usually found among grass tussocks near water and occasionally in a deserted penguin burrow. Though generally sited close to water, their nests have been found up to half a mile (1 km) inland.

FACT FILE

RANGE
Extreme S South America

HABITAT
Marine, especially rocky or gravelly coasts with offshore kelp beds

SIZE
24–33 in (61–84 cm)

TUFTED DUCK
Aythya fuligula

♀

♂

Tufted Ducks are gregarious birds that gather in large winter flocks and breed in loose colonies with their nests only a few feet apart. They have successfully moved into urban areas, colonizing park lakes and relying on people for scraps of food. Tufted Ducks are expert divers, able to reach depths of 16–20 ft (5–6 m) and remain underwater for 20–30 seconds. Their natural diet includes a wide variety of small mollusks and other aquatic invertebrates. The birds' nests are rarely far from water, concealed in tussocks of grass or under bushes. There, the females incubate the clutches of 7–12 eggs for about 24 days.

FACT FILE

RANGE
Eurasia

HABITAT
Lakes and ponds, sometimes in urban areas; winters on larger waters, estuaries, lagoons

SIZE
16–18½ in (40–47 cm)

COMMON EIDER
Somateria mollissima

♂ ♀ ♂ imm

FACT FILE

RANGE
Arctic and N temperate
Eurasia and North America

HABITAT
Coastal waters, estuaries

SIZE
19½–28 in (50–71 cm)

The down feathers that the female Common Eider plucks from her breast to line the nest have long been valued by people for insulating properties. In Iceland and Norway, breeding colonies of Common Eiders have been protected in order to maximize the production of eider down. Artificial nesting sites, the control of predators, and freedom from disturbance have enabled colonies of up to 10,000 pairs to become established and to be maintained over many years. The down is carefully collected and cleaned, and then sold to manufacturers of sleeping bags and comforters. Pure eider down is still considered superior to any man-made substitute and commands a high price. With heat loss delayed by the insulating feathers around the nest, the female incubates her clutch of 4–6 olive-green eggs for 28 days. After hatching, the young from several broods may come together to form a "crèche."

CANVASBACK
Aythya valisineria

The Canvasback is the largest of the pochards, or diving ducks. Both male and female have a relatively long sloping forehead, high, peaked crown, long bill, and long neck giving them a distinctive silhouette. The Canvasback is one of the typical breeding species of the prairie pothole country of southern Canada and northern U.S.A.—a vast area of rolling farmland with numerous small ponds (potholes) and marshes. The ducks feed mainly on aquatic vegetation, diving to depths of 30 ft (9 m). Often, they are closely followed by American Coots and American Wigeons. As soon as the Canvasbacks surface with a beak full of food, the other birds give chase and try to steal it.

FACT FILE

RANGE
S Canada to Mexico

HABITAT
Breeds on prairie marshes;
winters on lakes, lagoons
and estuaries

SIZE
19–24 in (48–61 cm)

Red-crested Pochard
Netta rufina

♀

♂

FACT FILE

RANGE
E Europe and SC Asia

HABITAT
Freshwater lakes, rivers,
deltas, and coastal lagoons

SIZE
21–22½ in (53–57 cm)

Red-crested Pochards have been spreading slowly north in Europe in the last 50 years. Their principle range is in the Mediterranean basin, but there are small pockets of breeding birds farther north in France, the Netherlands, and Denmark. Birds from these countries began to appear regularly in eastern England in the 1950s and 1960s, but the picture has become confused since then by escapes from captivity. This colorful species is popular with waterfowl keepers. The birds lay a clutch of up to 10 pale green or olive eggs in a well-concealed nest of aquatic vegetation, often built on a base of twigs. The female incubates alone, while the male moves away and joins other males for the annual molt.

BLACK (COMMON) SCOTER
Melanitta nigra

race *americana*
♂

race *nigra*
♂

race *nigra*
♀

The only relief from the all-black plumage of the male Black Scoter is a patch of yellow on the bill. In the Eurasian race *M. n. nigra*, the yellow covers only a small area halfway along the upper mandible, but in the North American race *M. n. americana*, the yellow covers most of the upper mandible and a prominent knob at the base of the bill. The latter is sometimes regarded as a separate species. During courtship, rival males rush at each other, seemingly skating over the surface with their heads outstretched, before coming to an abrupt halt in a flurry of spray. The females, too, will indulge in short rushes, usually at males they are rejecting. Another display given by the male involves raising the head and tail out of the water and producing a surprisingly pure and musical whistle. On a still day, the sound can carry for many hundreds of feet. Black Scoters nest beside large lakes as well as tiny moorland pools. If possible, they take their young to the sea soon after hatching.

FACT FILE

RANGE
Arctic and N temperate
Eurasia and North America;
winters to the S

HABITAT
Breeds on freshwater lakes
and marshes; winters in sea
bays and estuaries

SIZE
17–21 in (44–45 cm)

HARLEQUIN DUCK
Histrionicus histrionicus

FACT FILE

RANGE
Eurasia

HABITAT
Lakes and ponds, sometimes
in urban areas; winters on
larger waters, estuaries, lagoons

SIZE
16–18½ in (40–47 cm)

The strikingly marked Harlequin Duck is most at home in fast-flowing water, in which it is an expert swimmer. It can maneuver upstream with great skill, using eddies and slack water close to the bank, and sometimes rushes over the surface of the water, half-flying, half-swimming. Swimming with the current, it can pass through turbulent water without difficulty. When on the sea, which it mainly frequents in winter, it seems to spend most of the time just where the surf is breaking most vigorously. In flight, the Harlequin Duck passes fast and low over the water, precisely following each bend in a stream. Its nest, like that of most other ducks, is well concealed in vegetation, often overhanging a riverbank.

MARBLED TEAL
Marmaronetta angustirostris

The Marbled Teal takes its name from the whitish blotches that liberally mark its sandy-gray plumage. Though it resembles the teals, which are dabbling ducks, this species is now generally regarded as more closely related to the pochards, or diving ducks, and is increasingly known as the Marbled Duck. In particular, its courtship displays and other behavior patterns are more similar to the diving ducks. Breeding usually takes place in small colonies, sometimes with the nests no more than 3 ft (1 m) apart. The normal nest site is on the ground, concealed under low bushes or in thick vegetation close to the water. However, in southern Spain, nests have been found concealed in the roofs of grass-and reed-thatched huts. The birds do not excavate holes in the thatch but make use of natural cavities. Populations of this duck have declined over most of its range as shallow wetlands have been drained. However, there have been recent increases in numbers reported from Spain and North Africa.

FACT FILE

RANGE
Mediterranean region E to
Pakistan and NW India

HABITAT
Shallow freshwater and
brackish lakes, lagoons,
and marshes

SIZE
15–16½ in (39–42 cm)

RUDDY DUCK
Oxyura jamaicensis

race *jamaicensis*
♀

race *ferruginea*
"Peruvian Ruddy Duck"

race *jamaicensis*
♂

FACT FILE

RANGE
North and South America;
introduced into Britain

HABITAT
Breeds and winters on well-vegetated lakes; also winters on coastal lagoons and bays

SIZE
14–17 in (35–43 cm)

The male Ruddy Duck has white cheeks, a black cap, and a remarkable large blue bill, which gives the appearance of having just been painted. The tail consists of short stiff feathers that can be held up at a jaunty angle or kept flat along the water, contributing to the bird's dumpy appearance. The race *O. j. jamaicensis* occurs in the West Indies, while *O. j. ferruginea*, the Peruvian Ruddy Duck, is found in Peru and Bolivia. The male Ruddy Duck's display has been compared to the movements of a toy boat—he rushes over the water, paddling furiously, with his tail and lower body submerged, chest lifted out of the water, shoulders hunched, head down, and bill pressed down on to his raised chest. In another sequence, the male beats his bill rapidly on his chest before jerking his head forward and calling. Though the male does not help the female with incubation of the eggs, he often stays close by and then accompanies his mate and his growing brood.

OLDSQUAW
Clangula hyemalis

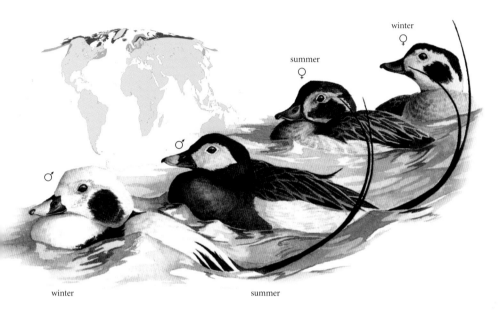

winter   summer

Oldsquaws have no fewer than 4 plumage changes during the course of a year. Both sexes have distinct summer, fall, winter, and eclipse plumages, the last one after breeding. They are sea ducks, spending nearly all of their time on salt water, though some birds breed beside freshwater. The courtship display of the male includes a far-carrying yodeling call that is wonderfully evocative of wild places. This duck is among the most accomplished divers of all wildfowl. It can remain underwater for as long as a minute and can reach depths of 180 ft (55 m). It pursues and catches fish during dives, while in shallower water, it plucks mollusks and crustaceans from the bottom.

FACT FILE

RANGE
Arctic Eurasia and North America; winters S to cool temperate regions

HABITAT
Breeds on tundra pools and by the coast; winters in sea bays

SIZE
14–18½ in (36–47 cm)

COMMON GOLDENEYE
Bucephala clangula

♀

♂

The Common Goldeneye's head is distinctively rounded and its steep forehead and short bill help identification at long range. The bird's natural breeding site is a cavity in a tree or a dead stump. Inside the nest-hole, the only lining materials for the 8–12 eggs are a few chips of rotten wood and some down feathers plucked from the female's breast. With forest clearance, tree holes have become rare, and the Common Goldeneye has declined in many areas. However, the species takes readily to nest-boxes and schemes involving several hundreds or even thousands of boxes are in operation. A nest-box scheme played a vital role in encouraging Common Goldeneyes to become established nesters in Scotland in the late 20th century.

FACT FILE

RANGE
N Eurasia and North America; winters to S

HABITAT
Breeds on lakes and pools in forests; winters on lakes, estuaries, sea bays

SIZE
16½–19½ in (42–50 cm)

RED-BREASTED MERGANSER
Mergus serrator

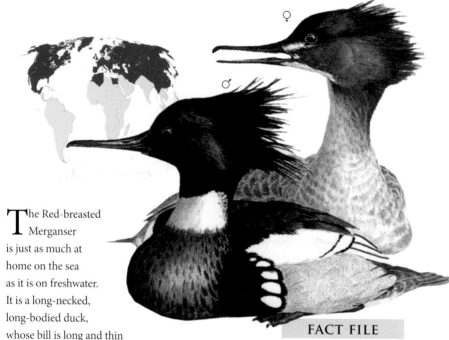

♀

♂

The Red-breasted
Merganser
is just as much at
home on the sea
as it is on freshwater.
It is a long-necked,
long-bodied duck,
whose bill is long and thin
with a hooked tip. Serrations along the
sides of the mandibles help the bird grasp fish.
The Red-breasted Merganser's appetite for fish has
brought it into conflict with fishing interests, particularly
on rivers with trout and salmon. The true effect of the
bird on fish stocks is mostly rather slight but, as is often
the case with wild predators, the bird is accused of
causing more damage than it actually carries out.
In some places, it is ruthlessly hunted.

FACT FILE

RANGE
N Eurasia and North
America; winters S to
Mediterranean, E China,
Gulf of Mexico

HABITAT
Breeds by rivers, estuaries,
and coasts; winters on
estuaries and coastal bays

SIZE
20½–23 in (52–58 cm)

HOODED MERGANSER
Mergus cucullatus

FACT FILE

RANGE
2 populations. W population breeds S Alaska to NW U.S.A.; winters S Alaska to California. E population breeds S Canada, N and C U.S.A.; winters Florida, N Mexico

HABITAT
Breeds and winters by rivers and lakes in forests; also winters on estuaries, coasts

SIZE
16½–19½ in (42–50 cm)

The male Hooded Merganser has a pronounced black and white crest on his head. When raised, the crest forms a quarter-circle of white outlined in black, greatly enlarging the appearance of the head. The bird's bill is long and thin and is used for grasping fish. The bird places its nest in a hole in a tree up to 52 ft (16 m) above the ground. The female lays and incubates 6–12 white eggs in the nest. It was once thought that the female brought the newly hatched young down to the ground in her bill. In fact, it has since been discovered that she calls encouragingly from the base of the tree and waits for the young to tumble down to her. Being so light and covered in soft down, they are rarely hurt in the process.

MUSK DUCK
Biziura lobata

♂

Both sexes of the Musk Duck have dark oily-looking plumage, but the male alone has a flat lobe of skin that hangs down below the lower mandible. The male is also nearly one-third larger than the female—an unusual size difference in any kind of bird. The male Musk Duck has a particular vigorous display—he throws his head back over his body, fans his tail back until it almost meets his head, and splashes noisily with both feet together. At the same time, he utters a loud grunt followed immediately by a piercing whistle. The male is strongly territorial, defending an area of freshwater against other males and attempting to attract several females to his territory. The bird's nest is concealed in vegetation at the water's edge or occasionally in a hollow of a tree stump or root.

FACT FILE

RANGE
S Australia and Tasmania

HABITAT
Breeds by freshwater and brackish lakes and marshes; winters on lakes, estuaries, and coastal bays

SIZE
Male: 24–29 in
(61–73 cm);
Female: 18½–23½ in
(47–60 cm)

African Fish Eagle
Haliaeetus vocifer
This eagle makes dramatic plunges into the water.

Stork-billed Kingfisher
Halcyon capensis
This kingfisher preys on a variety of creatures from a branch overhanging water.

Bearded Reedling
Panurus biarmicus
As its name suggests, this bird can be found among reedbeds.

Belted Kingfisher
Megaceryle alcyon
The Belted Kingfisher likes clear water and dives for fish.

MAKING A
LIVING BY
THE WATER

North American Dipper
Cinclus mexicanus
This bird lives in or beside
running water.

Several groups of birds include
species adapted to living with
water. Even birds of prey may
specialize in an aquatic way of life.
The White-tailed and Bald Eagles catch
fish from the surface. More dramatic
is the Osprey, which plunges headlong
and may briefly submerge in water, in
pursuit of fish. In Africa and Asia, fish
owls also adopt this hunting technique.

Other waterside specialists include
kingfishers, though worldwide, most members
of this family eat insects and some live in dry
places. Green and Belted Kingfishers are typical
plunge-divers seeking fish.

Many smaller birds live beside water, some of
them requiring water or waterside vegetation as
an essential part of their habitat, others equally
able to live in drier conditions, but find insect
food easy to come by in waterside locations.
Dippers are small songbirds that actually feed
underwater and swim on the surface—their
shape helps hold them underwater.

Grasshopper Warbler
Locustella naevia
The Grasshopper Warbler
can be found in marshes
and is known for its song.

OSPREY
Pandion haliaetus

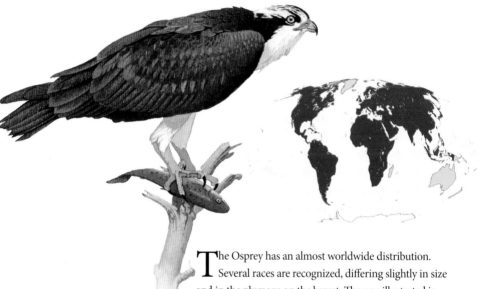

The Osprey has an almost worldwide distribution. Several races are recognized, differing slightly in size and in the plumage on the breast. The one illustrated is the North American race *P. h. carolinensis*, with a pale breast band (especially females) and a dark crown. The sexes sometimes differ in darkness of breast and crown, and females are larger than males, as in most birds of prey. Many Ospreys are coastal birds, but others breed around inland lakes and rivers. They feed almost exclusively on fish, soaring or circling up to 100 ft (30 m) above the water and often hovering briefly. When they catch sight of prey near the surface, they make a spectacular headlong plunge, throwing their feet forward at the last moment. The soles of the feet have sharp spines that help the birds grasp slippery fish. The large nest is built on a cliff, in a tree, or even on the ground in some habitats. Incubation of the 2–4 red-blotched creamy-white eggs takes about 38 days.

FACT FILE

RANGE
Breeds North America, Eurasia (mainly migrants), NE Africa, Australia; winter visitor and non-breeding migrant elsewhere

HABITAT
Coasts, rivers, lakes, wetlands

SIZE
21½–23 in (55–58 cm)

WHITE-TAILED SEA EAGLE
Haliaeetus albicilla

juv

White-tailed Sea Eagles are more bulky and have heavier bills than most other eagles likely to be seen in the same northerly areas. Their immense, broad wings give them a vulturelike appearance in flight. Magnificent soarers, White-tailed Sea Eagles can also perform deft aerial maneuvers if the occasion demands. They are skilled hunters of waterfowl, seabirds, and small mammals, and expertly snatch fish from the water's surface with their talons, usually after a shallow diving approach. They will rob birds, and perhaps even otters, of their prey and readily feed on carrion. The huge nest of sticks may be built in the top of a tree, on a crag, or on the ground on a small island if it is free of terrestrial predators.

FACT FILE

RANGE
W Greenland, Iceland, N, C, and SE Europe (recently reintroduced to Britain, to Rhum in the Inner Hebrides), N Asia

HABITAT
Coasts, lakes, rivers, wetlands

SIZE
27½–35½ in (70–90 cm)

African Fish Eagle
Haliaeetus vocifer

juv

With its unique combination of white, chestnut, and black plumage, this handsome fish eagle is unmistakable. It usually occurs close to water and is common over much of its range. It is a wonderfully agile flier and though it can lift fish straight from the surface while still in flight, it usually makes a spectacular plunge into the water with its feet stretched forward. The African Fish Eagle spends much of its time perching in an upright pose on a favorite lookout post. It is a vocal bird, both on its perch and while in flight. Its far-carrying, almost gull-like call is one of the best-known wild sounds of Africa.

FACT FILE

RANGE
Sub-Saharan Africa

HABITAT
Coastal and inland waters

SIZE
29–33 in (74–84 cm)

MARSH HARRIER
Circus aeruginosus

race *aeruginosus*

♀

♂

The harriers show markedly different plumage between the sexes, and this species is no exception. A widespread bird, it also shows marked variation across its geographical range. The race *C. a. aeruginosus* is from Europe, Israel, and Central Asia. Immatures are dark brownish in color. The Marsh Harrier generally nests in thick marsh vegetation, but it forages widely over nearby lowland habitats. Its prey ranges from small mammals and birds to amphibians, small fish, and large invertebrates. Carrion is sometimes eaten. When the female is incubating and tending the young, the male hunts for them both, calling his mate off the nest when he returns with food and dropping it for her to catch in midair.

FACT FILE

RANGE
W Europe E across Asia, Madagascar, Borneo, Australia

HABITAT
Usually lowland wetlands, especially with reedbeds

SIZE
19–23 in (48–58 cm)

GREATER SPOTTED EAGLE
Aquila clanga

juv

FACT FILE

RANGE
Breeds extreme E Europe E across N Asia; some birds resident, others winter in S of range, many winter as far S as NE Africa, Middle East, N India, and China

HABITAT
Mainly woodland or forest, usually near water or wetlands

SIZE
26–29 in (66–74 cm)

This thickset eagle's wings usually appear remarkably broad and blunt-ended in flight. Apart from a whitish crescent across the rump, adults are dark brown, though an uncommon pale phase is light buffish-brown. The prominent white spots on the juveniles give the species its name. Juvenile pale phase birds are strikingly pale sandy-buff with blackish flight feathers. The Greater Spotted Eagle scavenges widely, readily visiting carrion, but also hunts a wide variety of live prey, including small mammals, birds, reptiles, amphibians, and insects. Bird prey includes waterbirds, from young herons snatched from their nests to ducks and coots; in dealing with the latter, the eagle may separate one bird from a flock and then repeatedly stoop at it, forcing it to dive until it becomes exhausted and can easily be seized when at the surface.

STORK-BILLED KINGFISHER

Halcyon capensis

This is one of 3 species of stork-bill in Southeast Asia. They are among the biggest of the world's many kingfisher species and are exceeded in size only by the kookaburras, the Shovel-billed Kingfisher, and 3 giant *Megaceryle* species.

All species of stork-bill are vocal, but shy, birds. They are quite sluggish in habit, hunting from a branch overhanging water and plunging down vertically for prey, such as fish, frogs, crustaceans, insects, and small land vertebrates, such as lizards and nesting birds.

FACT FILE

RANGE
Southeast Asia, E Pakistan, India, Nepal, Sri Lanka

HABITAT
Rivers and streams in wooded lowlands, paddy fields; sometimes at wooded lakes or on the coast, especially among mangroves

SIZE
13–14 in (33–36 cm)

LESSER PIED KINGFISHER

Ceryle rudis

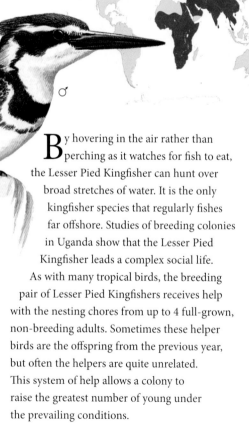

♀

♂

FACT FILE

RANGE
Africa, Middle East, S Asia

HABITAT
Lakes, broad rivers, estuaries

SIZE
10 in (25 cm)

By hovering in the air rather than perching as it watches for fish to eat, the Lesser Pied Kingfisher can hunt over broad stretches of water. It is the only kingfisher species that regularly fishes far offshore. Studies of breeding colonies in Uganda show that the Lesser Pied Kingfisher leads a complex social life.

As with many tropical birds, the breeding pair of Lesser Pied Kingfishers receives help with the nesting chores from up to 4 full-grown, non-breeding adults. Sometimes these helper birds are the offspring from the previous year, but often the helpers are quite unrelated. This system of help allows a colony to raise the greatest number of young under the prevailing conditions.

GREEN KINGFISHER
Chloroceryle americana

G enerally solitary, the Green Kingfisher could easily go unnoticed as it watches for prey but it has a characteristic habit of frequently raising its head and bobbing its tail, making it more visible. More than one Green Kingfisher may be seen in a suitable site, but they jealously guard their favorite hunting perches and are rarely seen together. Clarity of the water is important for hunting success. This species dives mainly for small fish that are 1–2 in (3–5 cm) in length. Its flight is straight and low over the water—the white on the wings and sides of the tail flashing conspicuously as it passes. This species utters a low, but distinctive *choot* or *chew* call and a descending series of *tew-tew-tew* notes. Its alarm call is a soft, tickling rattle.

FACT FILE

RANGE
S U.S.A. S to W Peru,
C Argentina, and Uruguay;
Trinidad and Tobago

HABITAT
Streams bordered by shrubby
habitat or forest, some
mountain stream

SIZE
7½ in (19 cm)

RIVER KINGFISHER
Alcedo atthis

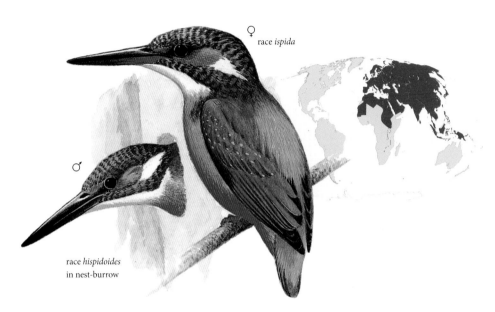

♀ race *ispida*

♂

race *hispidoides*
in nest-burrow

FACT FILE

RANGE
Breeds Europe, NW Africa,
Asia, Indonesia to Solomon
Islands; winters in S of range

HABITAT
Clear, slow-moving streams
and small rivers, canals,
ditches, reeds, marshes;
coasts in winter

SIZE
6 in (16 cm)

This bird and the Belted Kingfisher are the most
northerly breeding species of kingfishers. Along
with the European race *A. a. ispida*, the illustration shows
A. a. hispidoides, which ranges from northern Sulawesi
to the Bismarck Archipelago off Papua New Guinea.
The River Kingfisher is aggressively territorial and defends
a ½–3 ml (1–5 km) stretch of stream in winter, even from
its mate. In spring, it will drive away both rivals and small
songbirds. It excavates its burrow in the bank of a stream,
river, or gravel pit, usually above water. Most nest-burrows
are 18–36 in (45–90 cm) long. It rarely feeds away from
water and is an expert fisher.

Belted Kingfisher
Megaceryle alcyon

This solitary kingfisher prefers clear waters with overhanging trees, wires, or other perches. It usually hovers and dives for fish, but will also prey on insects, crayfish, frogs, snakes, and even the occasional mouse. A loud, dry, rattling call often announces the presence of the Belted Kingfisher. The breeding distribution of the bird seems to be limited only by the availability of foraging sites and earth banks in which it can excavate its nest chamber. The chamber lies at the end of an upward-sloping tunnel that is 3–16 ft (1–5 m) long.

FACT FILE

RANGE
Breeds through most of North America; winters in ice-free areas S to West Indies and N Colombia

HABITAT
Streams, rivers, ponds, marshes

SIZE
12 in (30 cm)

WHITE-THROATED DIPPER

Cinclus cinclus

juv

race *gularis*

The White-throated Dipper has a rich diet of insect larvae, fish fry and fish eggs, freshwater shrimp, and mollusks. Male and female both build the dome-shaped nest of moss, grass, and leaves, with an entrance directly over running water. The birds generally roost communally during the winter.

The race illustrated is *C. c. gularis*, from western and northern Britain. By contrast, the race *C. c. cinclus*, from northern Europe, has a black belly, while the race *C. c. leucogaster* of central Asia has a white belly.

FACT FILE

RANGE
Europe and C Asia

HABITAT
Fast-flowing upland streams

SIZE
7–8 in (18–21 cm)

NORTH AMERICAN DIPPER

Cinclus mexicanus

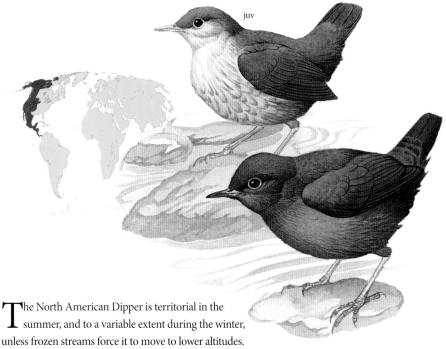

juv

The North American Dipper is territorial in the summer, and to a variable extent during the winter, unless frozen streams force it to move to lower altitudes. Like other dippers, it is entirely restricted to a life in and beside running water, which is exceptional for a small songbird. Dippers have a well-developed third eyelid, called a nictitating membrane, which protects its eyes from spray and when it is submerged. A dipper uses this membrane, which appears dramatically white against its dark head plumage, in a blinking action to signal alarm, excitement, or aggression, combining it with the bobbing, or "dipping," of its body.

FACT FILE

RANGE
Alaska to Mexico

HABITAT
Fast-flowing upland streams from 2,000 ft (600 m) up to the tree line

SIZE
7–8 in (18–21 cm)

SEDGE WREN
Cistothorus platensis

The Sedge Wren is slightly smaller than its relative the Marsh Wren *C. palustris*, but its song is quite distinct. The song is a weak, staccato, chattering trill, often delivered at night. The Sedge Wren's call note is a robust chip and is often doubled. As with other wrens, the male constructs dummy nests, from which the female Sedge Wren selects one for egg laying.

FACT FILE

RANGE
Breeds S Canada to E and C U.S.A., C Mexico S to Panama; winters in SE U.S.A., E Mexico

HABITAT
Wet, grassy meadows, sedge-dominated bogs and marshes

SIZE
4 in (11 cm)

BLUETHROAT
Luscinia svecica

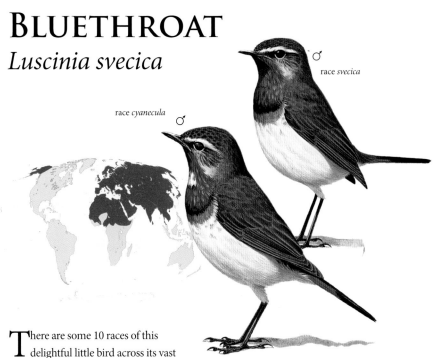

race *cyanecula* ♂

race *svecica* ♂

There are some 10 races of this delightful little bird across its vast range. In some, such as *L. s. svecica* of the northernmost latitudes, the center of the blue throat patch is red, while in others, such as *L. s. cyanecula* of Spain, central Europe, and parts of western U.S.S.R., the center spot is white. The Bluethroat flies close to the ground with a flitting action, followed by a flat glide or swoop into low cover. It has a more upright stance on the ground than the European Robin, but, like that species, it frequently flicks its tail. The song is more tinkling than that of the Nightingale. It may be uttered in the air or from a perch and often dominates the sounds of the Arctic tundra. The Bluethroat's diet consists largely of insects, sometimes caught in the air, but usually found by turning over leaf litter. The cup-shaped nest is built on the ground in thick vegetation, in a clump of grass, under a shrub, or sometimes in a hollow in a bank.

FACT FILE

RANGE
Eurasia, W Alaska

HABITAT
Wooded tundra, alpine meadows, dry, stony slopes, shrubby wetlands

SIZE
5½ in (14 cm)

BEARDED REEDLING
Panurus biarmicus

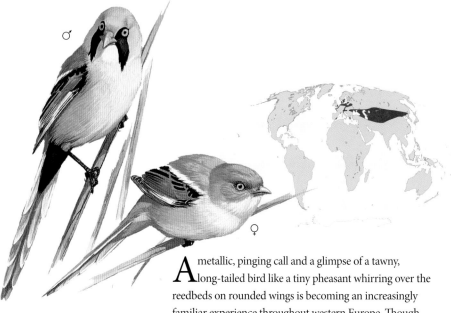

♂

♀

A metallic, pinging call and a glimpse of a tawny, long-tailed bird like a tiny pheasant whirring over the reedbeds on rounded wings is becoming an increasingly familiar experience throughout western Europe. Though they are basically sedentary birds, confined to a restricted habitat, Bearded Reedlings tend to flock together in the fall and will undertake mass movements away from their breeding quarters. Over the last 30 years or so these movements have become annual events and the species has colonized many sites beyond its previous range. Delightfully active and acrobatic, the Bearded Reedling is adept at straddling 2 reeds with its feet turned outward to grip each stem. In summer, it feeds mainly on insects, such as mayflies, but in winter, it will eat seeds, particularly those of the common reed *Phragmites*. Pairs often roost together on reed stems. The male is distinguished by his blue-gray head and jaunty black mustaches.

FACT FILE

RANGE
W Europe, Turkey, Iran
across Asia to E Manchuria

HABITAT
Reedbeds

SIZE
6 in (15 cm)

CETTI'S WARBLER
Cettia cetti

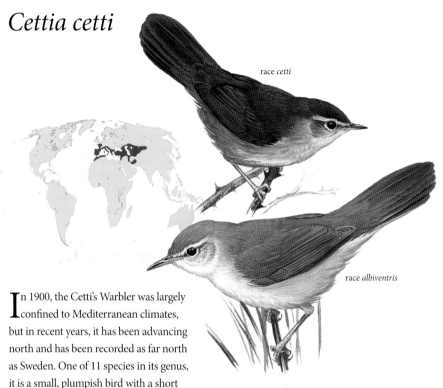

race *cetti*

race *albiventris*

In 1900, the Cetti's Warbler was largely confined to Mediterranean climates, but in recent years, it has been advancing north and has been recorded as far north as Sweden. One of 11 species in its genus, it is a small, plumpish bird with a short graduated tail; the males are heavier and longer-winged than the females. The central Asian race *C. c. albiventris* is larger and does not molt in spring like the western race *C. c. cetti* and as a result is much paler. Skulking and secretive, the Cetti's Warbler spends most of its time in dense cover but it occasionally shows itself on top of a bush or hedge. The position of the male is often given away by his brief, explosive song. Males play little part in breeding and may mate with 2 or more females. The female builds a nest in dense waterside vegetation or a low thicket and incubates the 3–5 brick-red eggs.

FACT FILE

RANGE
Mediterranean E to Iran and Turkestan; currently spreading N

HABITAT
Swamps, scrub alongside reedbeds, dense bushes, and hedges with brambles and tamarisks, edges of cornfields

SIZE
5½ in (14 cm)

GRASSHOPPER WARBLER
Locustella naevia

FACT FILE

RANGE
Europe S to N Spain and
Balkans, Baltic and W
U.S.S.R., C Asia E to Tian
Shan; winters NW Africa,
Iran, India, Afghanistan

HABITAT
Marshes, wet meadows with
shrubs, moist woods, osiers,
rough grassland, heaths,
dunes, conifer plantations

SIZE
5 in (12.5 cm)

Visually undistinguished, the Grasshopper Warbler is renowned for its extraordinary song, which is a vibrant, high-pitched mechanical trill resembling the whirr of an angler's reel and formed from double or triple notes produced at a rate of up to 1,400 triplets a minute. Tracking the source of the song can be difficult, for the bird seems able to "throw its voice" like a ventriloquist and it only rarely emerges from cover to sing in the open. Though a restless, agile species, it is very retiring—it generally forages for its insect prey deep within the foliage and, when disturbed, it often prefers to creep away through the grass rather than take flight. The nest is as hard to find as the bird itself, for it is usually well hidden in low vegetation with a concealed entrance.

Sedge Warbler
Acrocephalus schoenobaenus

The reed warblers of the genus *Acrocephalus* are essentially marsh-dwelling birds that occur in many parts of the Old World. The Sedge Warbler is one of the best known, for though it tends to keep to dense cover like most of its family, the male can be quite conspicuous early in the breeding season. It often advertises its presence with a stuttering call that is interspersed with mimicry of other species. Sedge Warblers feed mainly on slow-moving insects, which they pick out of low vegetation; in Africa, they take many lake flies. They nest in hedges, among osiers, reeds, or coarse grass, or even among standing crops of beans, rape, or cereals, binding the cup-shaped nest securely to the stems.

FACT FILE

RANGE
Europe (except Spain, Portugal, and some Mediterranean coasts) E to Siberia and SE to Iran; winters Africa S of Sahara and E of Nigeria

HABITAT
Osiers, marsh ditches, lakes, sewage works, gravel pits, conifer plantations, cereal, and rape fields

SIZE
5 in (12.5 cm)

Black-legged Kittiwake
Rissa tridactyla
This species is widespread,
in contrast to the Red-
legged Kittiwake.

Spotted Sandpiper
Actitis macularia
The Spotted Sandpiper
lives in a variety of habitats.

Common Snipe *Gallinago gallinago*
This bird will circle high in the air
and then dive downward
when displaying.

Beach Thick-knee
Esacus magnirostris
The Beach Thick-
knee captures its prey
while it walks along
the tide line.

SHOREBIRDS, GULLS, AND TERNS: THE GREAT TRAVELERS

Sooty Tern
Sterna fuscata
This tern's call
has earned it its
alternative name of
Wideawake Tern.

The shorebirds, or waders, include species that undertake vast migrations, breeding in the far north and wintering anywhere from temperate coasts to the southernmost tips of South America, Africa, and Australasia. The Turnstone is typical of beaches almost worldwide.

Other groups have different but remarkably similar species in different areas of the world—curlews, godwits, sandpipers, and oystercatchers all have familiar species in North America, Europe, Asia, Australia, and Africa, with each being slightly different from counterparts elsewhere.

Gulls and terns range widely, too, with fewer species and rather less variation, but American and European Herring Gulls, for example, are all but inseparable except in immature plumages.

Crab Plover
Dromas ardeola
This wader uses its
bill to break open
the shells of crabs.

GREATER PAINTED-SNIPE
Rostratula benghalensis

FACT FILE

RANGE
Africa, Asia

HABITAT
Well-vegetated freshwater
wetlands, including marshes
and paddy fields

SIZE
6½–9 in (17–23 cm)

When in the open, this wader freezes if danger threatens, holding its position until the threat has passed. The female is larger than the male and more brightly colored: the head, neck, and throat are a rich chestnut brown, and there is a distinct black band across the breast. She calls, a soft *koht-koht-koht*, at dusk and at night, often while in flight.

Males and females face each other in the courtship display, spreading their wings outward and forward until the tips are in front of their heads, at the same time fanning their tails. Each female may lay several clutches, each incubated by different males. The nest is a mound of waterweeds among vegetation in shallow water.

CRAB PLOVER
Dromas ardeola

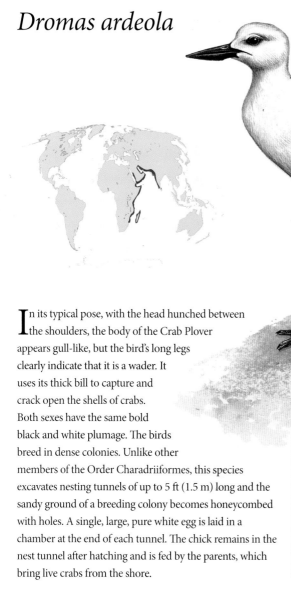

In its typical pose, with the head hunched between the shoulders, the body of the Crab Plover appears gull-like, but the bird's long legs clearly indicate that it is a wader. It uses its thick bill to capture and crack open the shells of crabs. Both sexes have the same bold black and white plumage. The birds breed in dense colonies. Unlike other members of the Order Charadriiformes, this species excavates nesting tunnels of up to 5 ft (1.5 m) long and the sandy ground of a breeding colony becomes honeycombed with holes. A single, large, pure white egg is laid in a chamber at the end of each tunnel. The chick remains in the nest tunnel after hatching and is fed by the parents, which bring live crabs from the shore.

FACT FILE

RANGE
Coasts of Indian Ocean;
breeds Gulf of Oman, Gulf
of Aden, and Red Sea

HABITAT
Sandy beaches and
coastal dunes

SIZE
15–16 in (38–41 cm)

EURASIAN OYSTERCATCHER
Haematopus ostralegus

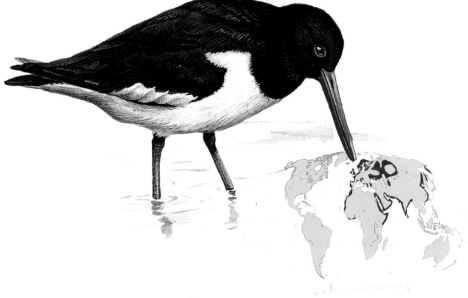

FACT FILE

RANGE
Breeds Eurasia; winters S to
Africa and Indian Ocean

HABITAT
Breeds on coasts; near inland
fresh waters; winters on coasts

SIZE
16–18 in (40–46 cm)

Like that of other oystercatchers, the stout orange bill of the Eurasian Oystercatcher is triangular in cross-section and reinforced so that it does not bend easily. Birds that feed mainly on cockles and mussels have a bladelike tip suited to hammering open the shells and severing the muscle within. Those that feed on invertebrates in sand and mud have finer, more pointed bills for probing.

Territories are advertised and boundaries disputed by noisy displays involving loud, piping calls, which take place among groups of these birds all year round. The nest is a mere scrape on the ground, sometimes decorated with stones or pieces of shell. Outside the breeding season the adults develop a white neck collar, making them similar to an immature bird in its first winter.

AMERICAN BLACK OYSTERCATCHER

Haematopus bachmani

Oystercatchers have loud, piercing calls that can be heard above the sound of crashing surf, facilitating communication between individuals. They are noisy birds, frequently calling at night. The downy, highly mobile chicks follow their parents, which first present food to them and then train them to find and procure food for themselves. Opening mollusks and crab shells is a very difficult task for the bird; it can be accomplished in less than a minute by an experienced bird but many months are required to master the technique. Young birds are able to fly in about 35 days but associate with their parents for up to a year, learning feeding techniques.

FACT FILE

RANGE
Pacific coast of North America, from Alaska to Baja California

HABITAT
Rocky coasts and islands, occasionally sandy beaches

SIZE
15 in (38 cm)

PIED AVOCET
Recurvirostra avosetta

chick

The delicately pointed, upturned bill of the Pied Avocet is a highly specialized feeding tool. It is held slightly open just below the surface of shallow brackish water or very soft mud and swept from side to side as the bird walks slowly forward. Small invertebrates and larger worms are located by touch. The species can also upend in deeper water and sweep its bill to and fro through the mud on the bottom. Pies Avocets nest on open ground on muddy islets, forming dense colonies that usually contain 10–200 pairs. The female lays 3–5 eggs in a lined scrape or on top of a mound of vegetation. Each pair defends the few yards around the nest against rivals, but all the pairs cooperate in driving away intruders. The call of the Pied Avocet is a loud *klute-klute-klute*.

FACT FILE

RANGE
Breeds W and C Eurasia;
winters Africa, Middle East

HABITAT
Saline or brackish wetlands,
coastal and inland

SIZE
8 in (20 cm)

BLACK-WINGED STILT
Himantopus himantopus

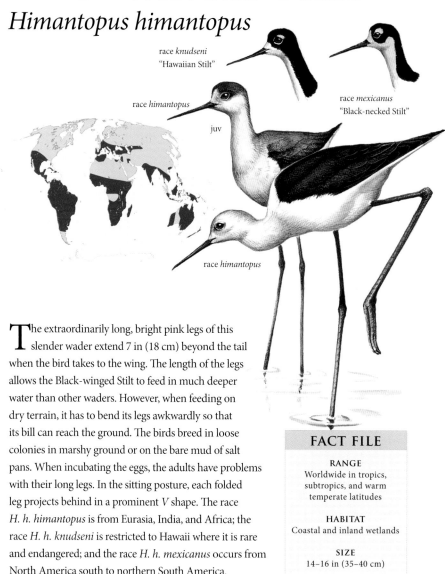

race *knudseni*
"Hawaiian Stilt"

race *himantopus*

race *mexicanus*
"Black-necked Stilt"

juv

race *himantopus*

The extraordinarily long, bright pink legs of this slender wader extend 7 in (18 cm) beyond the tail when the bird takes to the wing. The length of the legs allows the Black-winged Stilt to feed in much deeper water than other waders. However, when feeding on dry terrain, it has to bend its legs awkwardly so that its bill can reach the ground. The birds breed in loose colonies in marshy ground or on the bare mud of salt pans. When incubating the eggs, the adults have problems with their long legs. In the sitting posture, each folded leg projects behind in a prominent *V* shape. The race *H. h. himantopus* is from Eurasia, India, and Africa; the race *H. h. knudseni* is restricted to Hawaii where it is rare and endangered; and the race *H. h. mexicanus* occurs from North America south to northern South America.

FACT FILE

RANGE
Worldwide in tropics, subtropics, and warm temperate latitudes

HABITAT
Coastal and inland wetlands

SIZE
14–16 in (35–40 cm)

BEACH THICK-KNEE
Esacus magnirostris

FACT FILE

RANGE
Coasts of N Australia, New
Guinea, New Caledonia,
Solomon Islands,
Philippines, Indonesia,
Andaman Islands

HABITAT
Sandy beaches and
inshore reefs

SIZE
21–23 in (53–58 cm)

Also known as the Beach Stone-curlew, this species is
a plain-colored, large, and heavily built thick-knee
that wails mournfully at night. It is almost exclusively a
ground-dweller, running to escape danger, and only taking
to the air when hard pressed. The Beach Thick-knee feeds
on crabs and other hard-shelled marine animals by seizing
them as it walks along the tide line with its characteristic
dawdling gait. It uses its heavy bill to crack and smash the
shells of its prey open. Feeding takes place mostly at night.
The bird rests during the day by squatting on its long legs.
Its nest is a scrape in shingle or sand just above the high
water mark, placed among beach debris or, on occasion,
under a sheltering bush.

Egyptian Plover
Pluvianus aegyptius

The Egyptian Plover's breeding habits are unique among waders. The adults bury their 2–3 eggs under a thin layer of sand, about ⅛ in (3 mm) thick. Warmed by the sun, the sand helps to incubate the clutch. Should the weather become cooler, the parents incubate the eggs by sitting on them in the normal way, and they also do so at night. Conversely, if the daytime temperature becomes very high, the male or female will cool the eggs by shading them with its body or by wetting the sand with water carried in its belly feathers. The young hatch in the sand and quickly leave the "nest" area. Until the chicks are about 3 weeks old, they crouch whenever danger threatens and the adults quickly cover them with sand.

FACT FILE

RANGE
W, C, and NE Africa

HABITAT
River valleys and marshes

SIZE
7½–8 in (19–21 cm)

Northern Lapwing

Vanellus vanellus

FACT FILE

RANGE
Breeds temperate Eurasia;
most populations migrate S
in winter to Mediterranean,
India, China

HABITAT
Breeds on open ground,
including farmland,
freshwater marshes,
salt marshes; also visits
estuarine mudflats

SIZE
11–12 in (28–31 cm)

The Northern Lapwing is a common and familiar bird over much of Europe. Its broad wings produce a distinctive creaking sound during the wheeling, rolling, and tumbling display flight. The sounds are accompanied by a characteristic plaintive cry, from which the bird takes its alternative English name, Peewit. The female carries out most of the incubation of the 4 eggs. The young are perfectly camouflaged, and crouch down the instant they hear their parents' alarm calls. Intensive farming and land drainage have drastically reduced the bird's habitat in many areas and the species is also vulnerable to severe winters. In many areas of north-west Europe, the birds are now found nesting only in specially protected nature reserves with ideal habitat management.

AMERICAN GOLDEN PLOVER
Pluvialis dominica

non-breeding

♂

There are 3 species of "golden plovers" in the northern hemisphere: this one in the New World, the Pacific Golden Plover *P. fulva* of Asia, and the European Golden Plover *P. apricaria*. All have black on the face and underparts in summer, recalling the pattern of the larger Gray Plover *P. squatarola*. The European species has white underwings; the others have gray. There is a striking difference between the breeding and non-breeding plumage. However, the upperparts of adult and immature have a golden hue all year. The speckling on the back camouflages the incubating bird. This species undertakes a long seasonal migration apparently accomplished non-stop by many birds. Eastern Canadian breeders take the shortest "Great Circle" route over the western Atlantic, but other populations take inland routes across the U.S.A.

FACT FILE

RANGE
Breeds Arctic North America; winters C South America

HABITAT
Breeds on tundra; otherwise coastal mudflats, inland marshes, and grassland

SIZE
9½–11 in (24–28 cm)

KILLDEER
Charadrius vociferus

FACT FILE

RANGE
Breeds S Canada to SC Mexico, West Indies, coastal Peru, and extreme N Chile; winters N South America

HABITAT
Open expanses near wet areas, ponds, rivers

SIZE
8–10 in (21–25 cm)

This plover takes its name from its *killdee-killdee* calls. Widespread through the Americas, it is a rare visitor to Europe and Hawaii. Killdeers make their nest scrapes in an area with short, sparse vegetation or none at all. Bits of grass, pebbles, and other loose material end up in the scrape as a result of a ritual ceremony in which the birds slowly walk away from the nest while tossing the material over their shoulders. The birds often nest on gravel streets where their eggs are crushed by cars. They may also nest on flat gravel rooftops, though the chicks are sometimes trapped by parapets or perish when they leap to the ground. However, they often survive leaps from buildings as much as 3–4 stories high.

BLACK-BELLIED (GRAY) PLOVER
Pluvialis squatarola

non-breeding

♂

Though the black axillary feathers, or "armpits," of this bird are seen only in flight, they form a unique and distinctive field character. They are especially useful in winter when the overall gray plumage of the standing bird lacks any obvious distinguishing marks. The Black-bellied Plover is highly migratory, reaching the southern shores of Australia and South Africa from its breeding grounds in the high Arctic. Adults share incubation of the 4 eggs, but the male takes more responsibility for rearing the chicks.

FACT FILE

RANGE
Breeds circumpolar Arctic; winters worldwide on coasts to S

HABITAT
Breeds on lowland tundra; winters on coastal mudflats and lagoons

SIZE
10½–12 in (27–30 cm)

RINGED PLOVER
Charadrius hiaticula

FACT FILE

RANGE
Breeds Arctic E Canada
and Eurasia; winters
Europe and Africa

HABITAT
Breeds on shingle beaches
and tundra; winters along
coasts and some
inland wetlands

SIZE
7–8 in (18–20 cm)

The white collar and broad black breast band of the adult Ringed Plover gives the species its name. The breast band is blacker and deeper in the adult male than in the female. This species' wistful *too-li* call is used by the off-duty bird to call its mate from the nest when danger threatens. Later, when the chicks have hatched, Ringed Plovers are experts at the broken-wing display, in which they pretend to be injured. They flutter with drooped wings and spread tail while calling loudly and plaintively in an effort to distract a potential predator away from the vulnerable young in the nest.

BAR-TAILED GODWIT
Limosa lapponica

non-breeding

♂

The rich mahogany-chestnut summer plumage of this species provides good camouflage for nesting birds against the colored mosses and low scrub of the Arctic tundra. The thin, slightly upturned bill is approximately 20 percent longer in the female than in the male. In a typical flock feeding along the edge of the tide, the females wade and probe in deeper water than the males. The nest is usually placed on a drier raised hummock above the damp ground. Both sexes incubate the 4 eggs and rear the young—the off-duty bird often perching, incongruously, on top of a bush or low tree.

FACT FILE

RANGE
Breeds Eurasian Arctic; winters S to S Africa and Australasia

HABITAT
Breeds on marshy and scrubby tundra; winters on estuaries

SIZE
14½–16 in (37–41 cm)

GREATER YELLOWLEGS
Tringa melanoleuca

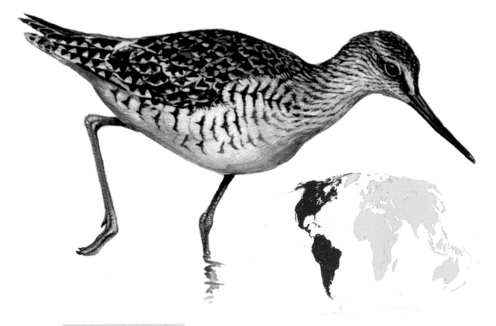

FACT FILE

RANGE
Breeds N North America,
non-breeders S along coasts;
winters SW Canada to
extreme S South America;
vagrant elsewhere

HABITAT
Breeds on tundra;
otherwise coastal mudflats
and marshes, shores of
inland lakes

SIZE
14 in (36 cm)

The Greater Yellowlegs is usually seen alone or in small flocks. It nests in muskeg country, along the tundra-forest edge and on the high tundra of North America, migrating south for the winter to warmer regions. It feeds mainly in shallow waters, wading in to sweep its bill with a sideways motion or dash about after visible prey. Its diet consists largely of small fish, insects and their larvae, snails, worms, and tadpoles. This species sometimes associates with the Lesser Yellowlegs *T. flavipes*, especially during migration. When it does, its greater size and its longer, stouter bill make it easy to distinguish.

SPOTTED SANDPIPER
Actitis macularia

The Spotted Sandpiper occupies varied habitats from sea level to the tree line. Tipping up its tail at nearly every step, it teeters along the water margins, stalking diverse aquatic and terrestrial insects. Its main call is a clear, whistled *peet-weet* and it utters a repetitious *weet* during the breeding season. Its curious, jerky flight, with short glides on down-bowed wings alternating with groups of shallow, flickering wing beats, is shared only by its close relative in the Old World, the Common Sandpiper *A. hypoleucos*. Like a number of other American waders, including the Greater and Lesser Yellowlegs, the Spotted Sandpiper is a rare wanderer to Europe; in 1975, a pair nested in Scotland, but they deserted their 4 eggs. Certain females of this species may be polyandrous, mating with 2 or more males. Females may or may not help with incubation.

FACT FILE

RANGE
Breeds North America; winters S from Mexico to S Brazil

HABITAT
Open and wooded areas, usually near water

SIZE
7½ in (19 cm)

RUDDY TURNSTONE
Arenaria interpres

non-breeding

♂

FACT FILE

RANGE
Breeds circumpolar Arctic,
and N temperate region
in Scandinavia; winters S
worldwide to C Argentina,
South Africa, and Australia

HABITAT
Breeds on tundra
and coastal plains;
winters on coasts

SIZE
8–10 in (21–26 cm)

Flocks of these birds walk busily over heaps of seaweed on the shore, flicking pieces with the bill and quickly grabbing any morsel of food that appears. Invertebrates and their larvae make up the main part of their diet, but the birds also feed on other items, including dead animals and human waste. Males are more dull than females. During the breeding season, each pair defends a small territory around the nest site. Both adults share in incubating the 3–4 eggs and in rearing the young.

WESTERN CURLEW
Numenius arquata

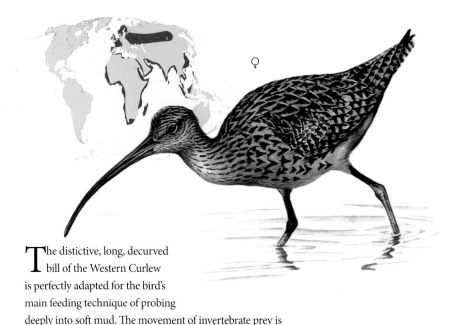

♀

The distictive, long, decurved bill of the Western Curlew is perfectly adapted for the bird's main feeding technique of probing deeply into soft mud. The movement of invertebrate prey is detected by the sensitive tips of the mandibles. Food items are also picked from the surface, including mollusks, worms, and, while inland, even berries.

The only difference in the appearance of males and females is that the bill is longest in the adult female. Its non-breeding call, for which the bird is named, is a mournful-sounding *curloo-oo*. During courtship, the male Western Curlew indulges in a shallow, gliding flight accompanied by a rich, bubbling and trilling song, which can also be heard outside the breeding season. It is one of the most evocative spring sounds of upland moorland areas. The nest is a shallow hollow among grass tussocks, lined with grass stems.

FACT FILE

RANGE
Breeds temperate and N Eurasia, winters S to Africa, India, SE Asia

HABITAT
Breeds on open vegetated land; winters on coastal wetlands

SIZE
20–24 in (50–60 cm)

Red-necked Phalarope

Phalaropus lobatus

non-breeding

♀

Like the other 2 species of phalarope, Red-necked Phalaropes are adapted for swimming, rather than merely wading, with lobed toes to propel them through water. They feed daintily while swimming, spinning around to stir up the water and then picking small creatures from the surface. As with the other phalaropes, the female Red-necked Phalarope is more brightly colored than the male and takes the dominant role in courtship. After she has laid her clutch of 4 eggs in a shallow cup among moss, the male carries out all the incubation and rearing of the young— hence the adaptive significance of his duller plumage, since it is an advantage for him to be better camouflaged than the female. A proportion of females lay further clutches for other males, but still take no part in caring for the young.

FACT FILE

RANGE
Breeds circumpolar Arctic and N temperate regions; winters in the tropics

HABITAT
Breeds on freshwater marshes, including uplands; marine in winter

SIZE
7–7½ in (18–19 cm)

RED KNOT
Calidris canutus

non-breeding

The Red Knot undergoes a marked change in plumage through the year, from its rich chestnut-orange breeding dress to the pale gray of winter. Fall and winter flocks can be tens of thousands of birds strong and will perform rapid aerial maneuvers in which the entire flock twists and turns with great synchrony. At a distance, the flock can give the appearance of a writhing mass of smoke. Though the rather more dull, white-bellied female helps with incubation, she departs soon after, leaving the male to rear the brood. This means that more invertebrate food is available for the chicks than if a second adult were present.

FACT FILE

RANGE
Breeds Arctic Canada and Siberia; winters South America, S Africa, and Australia

HABITAT
Breeds on tundra; winters on estuaries and beaches

SIZE
9–10 in (23–25 cm)

Ruff
Philomachus pugnax

♂

♀

FACT FILE

RANGE
Breeds N Eurasia; winters
Mediterranean, S Africa,
India, and Australia

HABITAT
Lowland freshwater
wetlands, marshes,
and wet pastures

SIZE
8–12½ in (20–32 cm)

In the breeding season, the male Ruff develops remarkable plumes, variable in both color and pattern, around his head and neck. Groups of rival males form "leks" at which they display to one another and advertise to visiting females. Some hold small territories within the lek, while others move from one position to another. Each female may mate with one or more males before moving up to 1,600 ft (500 m) away from the lek area. There, she makes a shallow scrape, lays a clutch of 4 eggs, incubates them, and rears the young. The males take no part in nesting.

COMMON SNIPE
Gallinago gallinago

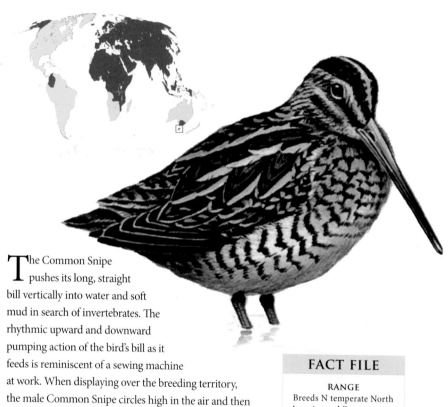

The Common Snipe pushes its long, straight bill vertically into water and soft mud in search of invertebrates. The rhythmic upward and downward pumping action of the bird's bill as it feeds is reminiscent of a sewing machine at work. When displaying over the breeding territory, the male Common Snipe circles high in the air and then dives groundward. As he does so, he holds out his outer tail feathers at a sharp angle to the rest of his tail. The air rushing past makes the feathers vibrate, giving out a bleating sound, often known as "drumming." Other species of snipes make a similar noise with their tails. The nest is a scrape, often placed near a grassy tussock.

FACT FILE

RANGE
Breeds N temperate North America and Eurasia; many winter widely to S

HABITAT
Freshwater and coastal wetlands

SIZE
10–10½ in (25–27 cm)

DUNLIN
Calidris alpina

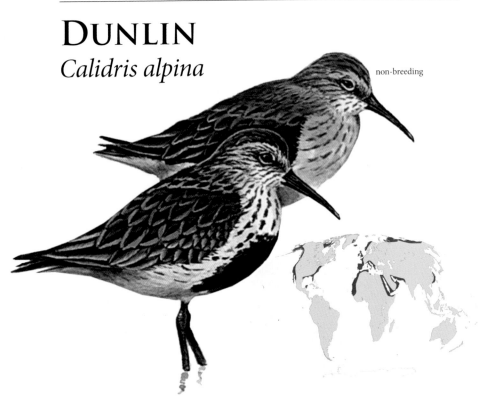

non-breeding

FACT FILE

RANGE
Circumpolar Arctic and N temperate regions; winters S to subtropics

HABITAT
Breeds on tundra, moorland, marshes; winters on estuaries or inland wetlands

SIZE
6–8½ in (16–22 cm)

The Dunlin is the most common small wader of the northern hemisphere. It may occur in passage and winter flocks of tens of thousands in favored estuaries, but also appears in tiny groups in small coastal creeks. Extensive ringing has shown that wintering birds return year after year not just to the same estuary but to the same part of the estuary, with almost no mixing between flocks only a few miles apart. This species is very variable in size with birds of southern and western races being far smaller than those of some of the east Siberian and New World races. Dunlins nest in shallow depressions among short vegetation. Both sexes share in the incubation of the 4 eggs, but the female leaves soon after the young have hatched.

SPOON-BILLED SANDPIPER
Eurynorhynchus pygmeus

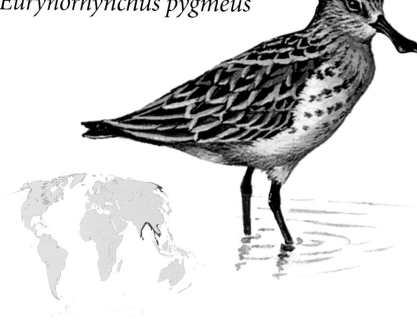

The spoon-shaped bill of this species is an adaptation for feeding. The bird forages in soft mud or shallow water, sweeping its bill from side to side as it walks. The broadened bill tip is already obvious in the newly hatched young. The Spoon-billed Sandpiper's total population is small, with a best estimate of no more than 2,800 pairs; the true figure may well be considerably lower. However, accurate censuses are extremely difficult to carry out. The male apparently carries out most of the incubation of the 4 eggs as well as rearing the young.

FACT FILE

RANGE
Breeds NE Siberia; winters coasts of SE India to Burma, coastal SE China

HABITAT
Breeds on coastal marshy tundra; winters on muddy coasts and brackish lagoons

SIZE
5½–6 in (14–16 cm)

LONG-TAILED JAEGER
Stercorarius longicaudus

typical

The Long-tailed Jaeger, or Long-tailed Skua, is the most slender and elegant member of its family. It has a central pair of tail feathers that extend about 7 in (18 cm) beyond the remainder. During the breeding season, this species concentrates on hunting lemmings, rather than parasitizing other seabirds. The breeding performance of the jaegers is closely linked to the abundance of lemmings. In years when lemmings are scarce, the jaegers may not breed at all, or will fail to rear any young, but they breed very successfully in good lemming years. The nest is a shallow, lined scrape on the ground, and both parents incubate the 2 eggs. When food is short, the first chick to hatch may kill the second chick so that it does not have to share the limited food supply.

FACT FILE

RANGE
Breeds circumpolar high
Arctic; winters in
S hemisphere

HABITAT
Breeds on Arctic tundra and
marshes; winters at sea

SIZE
20–23 in (50–58 cm)

Mew (Common) Gull
Larus canus

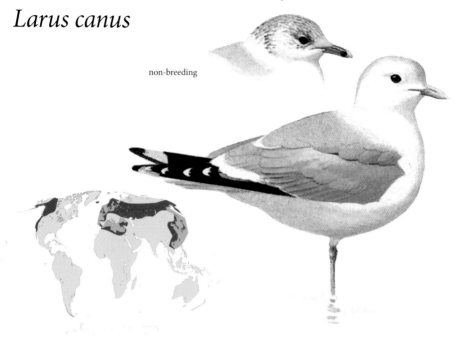

non-breeding

Mew Gulls nest in colonies, sometimes on the ground, but often in the tops of bushes or scrub. In parts of Norway, the birds are encouraged to nest in open boxes on poles, because their eggs provide food for local inhabitants. The usual clutch contains 3 eggs, which are incubated by both parents for about 3 weeks. The young Mew Gulls take a further 4–5 weeks to fledge. In winter, the birds gather in large roosts on estuarine mudflats or large freshwaters. They may feed along the shore or fly well inland to forage for invertebrates in pastures, returning to the roosting site in the evening.

FACT FILE

RANGE
Temperate and N Eurasia and NW North America

HABITAT
Breeds in marshes and wet scrubland; winters on the shores of seas and large lakes

SIZE
16–18 in (40–46 cm)

HERRING GULL
Larus argentatus

race *argenteus*

The Herring Gull is one of the most successful of all seabirds, having adapted in recent decades to living on human refuse and nesting on roofs in many areas. It is now commonplace in urban areas, some well inland, and strenuous efforts have been made to prevent the gulls from nesting on, and fouling, important buildings. The Herring Gull's nest is an untidy gathering of vegetation and scraps of garbage. Soon after they have hatched in the nest, the young Herring Gulls peck instinctively at the prominent red spot on the lower mandible of their parents' bills. The action induces an adult bird to regurgitate the food in its crop. The race *L. a. argenteus* breeds in northwest Europe, while the yellow-legged *L. a. michahellis* breeds in the Mediterranean region, though it is expanding northward.

FACT FILE

RANGE
Circumpolar N, temperate and Mediterranean

HABITAT
Varied; coastal and inland, including urban areas

SIZE
22–26 in (56–66 cm)

GREAT BLACK-BACKED GULL

Larus marinus

The Great Black-backed Gull takes its name from its large size and black back and wings. It is a highly predatory gull, especially in the breeding season when it may live almost exclusively on seabirds and, if they are available, rabbits. At other times of the year, it often obtains food by scavenging at garbage tips, around fishing ports, and along the shore. Breeding may take place in loose colonies or the birds may build solitary nests, usually on a rock promontory or on a ridge with a good view all around. The nest is a bulky pile of vegetation.

FACT FILE

RANGE
Coastal NE North
America, SW Greenland,
and NW Europe

HABITAT
Coasts, locally inland
waters and moors

SIZE
28–31 in (71–79 cm)

ROSS'S GULL
Rhodostethia rosea

non-breeding

FACT FILE

RANGE
Arctic Siberia, Canada,
and Greenland

HABITAT
Breeds on Arctic tundra and
scrub; winters at sea

SIZE
12–12½ in (30–32 cm)

Perhaps the most beautiful of all the gulls, Ross's Gull in its breeding plumage has white underparts suffused with pink. For a long time, the breeding grounds of this small gull were quite unknown, though they were obviously in northern latitudes. They were finally discovered in northern Siberia in 1905, where the birds were found nesting among low scrub in river valleys. More recently, breeding has been confirmed in northern Canada and in Greenland. The birds build a nest of vegetation that may stand 8 in (20 cm) above the damp, boggy ground.

SWALLOW-TAILED GULL
Creagrus furcatus

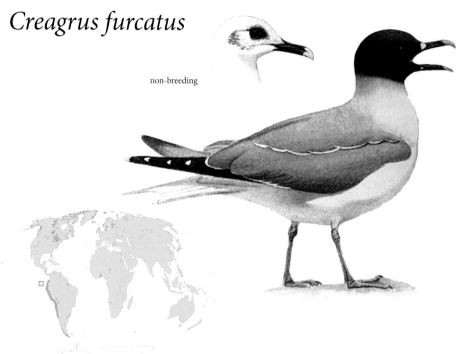

non-breeding

With its dark gray head, boldly patterned wings, and deeply forked tail, the Swallow-tailed Gull has a particularly striking appearance. Its eyes are much larger than those of most gulls and are more forward-facing, giving the bird binocular vision. These are adaptations that help the birds find food at night. Swallow-tailed Gulls range several hundred miles from their breeding grounds out over the ocean and take squid and other sea creatures that come to the surface after dark. The bird's nest is a cup of small stones among rocks in which it lays a single egg. The incubation period is 34 days and the young chick is then brooded and guarded by one parent throughout the day.

FACT FILE

RANGE
Breeds Galapagos Islands
and Colombia; winters off
NW South America

HABITAT
Breeds on coasts;
winters on open sea

SIZE
21½–23½ in (55–60 cm)

BLACK-LEGGED KITTIWAKE
Rissa tridactyla

first winter

FACT FILE

RANGE
Circumpolar Arctic and temperate coasts

HABITAT
Coastal and open sea

SIZE
15–18 in (39–46 cm)

This widespread species is often known simply as the Kittiwake: the other species, the Red-legged Kittiwake, is restricted to the Bering Sea region of the north Pacific. Kittiwakes traditionally nest in large colonies on narrow cliff ledges, building a nest of seaweed, lined with grass. As the seaweed dries, it cements itself to the rock. In recent years, ledges on buildings have proved attractive to Black-legged Kittiwakes, and there are now many nesting colonies around harbors and fishing ports. The 2–3 eggs are incubated by both parents for about 27 days. The young, when they hatch, have strong claws on their toes, enabling them to cling to their precarious home.

SABINE'S GULL
Xema sabini

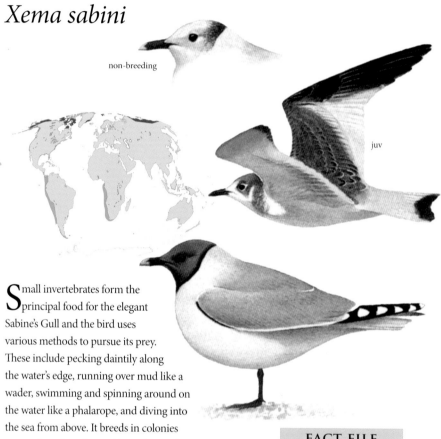

non-breeding

juv

Small invertebrates form the principal food for the elegant Sabine's Gull and the bird uses various methods to pursue its prey. These include pecking daintily along the water's edge, running over mud like a wader, swimming and spinning around on the water like a phalarope, and diving into the sea from above. It breeds in colonies on coasts and tundra, and it steals eggs from the nesting colonies of Arctic Terns. It makes a ground nest lined with grass, in which it lays 2–3 olive-brown eggs. Presumably as a defense against predators, parents lead their young away from the nest as soon as they hatch. The adults also perform a distraction display if danger threatens their chicks. Both these types of behavior are more typical of waders than of gulls.

FACT FILE

RANGE
Circumpolar Arctic

HABITAT
Breeds Arctic coasts
and islands; winters
on the open sea

SIZE
13–14 in (33–36 cm)

BLACK TERN
Chlidonias nigra

non-breeding

FACT FILE

RANGE
Breeds central S Eurasia and North America; winters in tropics mainly N of Equator

HABITAT
Breeds on marshes and pools; winters on coasts

SIZE
8½–9½ in (22–24 cm)

The Black Tern feeds mainly on aquatic insects and their larvae and obtains almost all of its prey in flight. Instead of hovering above the water and then diving into it, like many other species of terns, the Black Tern and the other "marsh terns" of the genus *Chlidonias* hover and then swoop down to peck from the water's surface in mid-flight. Black Terns also hawk for insects in midair and will even follow a plow on farmland, darting down to snatch small worms and other invertebrates exposed on the soil. The bird's nest is composed of aquatic vegetation and is often floating, anchored to growing plants. The usual clutch is 3 eggs. The young leave the nest when they are a few days old and hide in nearby vegetation.

CASPIAN TERN
Hydroprogne caspia

The Caspian Tern is the largest of the terns, with an orange-red, daggerlike bill. It breed in large colonies or in solitary pairs. Its nests are merely scrapes in the ground, sometimes with a little vegetation. The Caspian Tern's range is remarkable in covering all 5 continents, but the breeding areas are hundreds of miles apart. This is probably because the species requires remote coastal and marshland nesting sites with rich feeding areas close by. Such places are naturally scarce and some potential sites may have become unsuitable due to human activity.

FACT FILE

RANGE
Worldwide, but breeding range is fragmentary

HABITAT
Breeds on coastal and inland marshes; winters on coasts

SIZE
19–23 in (48–59 cm)

ARCTIC TERN
Sterna paradisaea

FACT FILE

RANGE
Breeds circumpolar Arctic
and sub-Arctic; winters S
to Antarctic Ocean

HABITAT
Breeds on coasts;
winters at sea

SIZE
13–15 in (33–38 cm)

The annual migration of the Arctic Tern is probably the longest of any bird. From their breeding grounds in the Arctic, the birds fly south to winter in the seas just north of the Antarctic pack-ice, a round trip of more than 20,000 mls (32,000 km) every year. Because they breed in the northern summer and visit the Antarctic Ocean during the southern summer, they experience more daylight hours in the course of a year than any other living animal. Arctic Terns breed in colonies on shingle beaches and among rocks. Like some other terns, they vigorously defend their nests against predators, or even approaching humans, by dive-bombing and actually striking the intruders around the head with their beaks.

ROYAL TERN
Thalasseus maximus

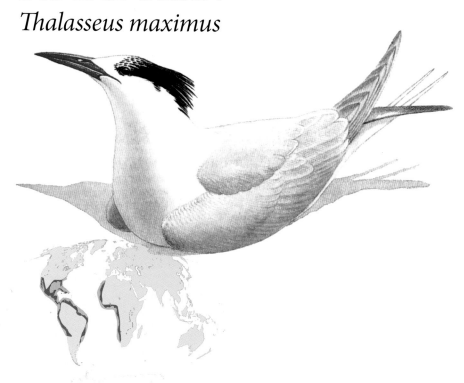

In the past, Royal Terns were commonly observed on the
west coast of Africa but it was thought that they were
migrants from the American breeding range. Then, a large
colony was discovered during the 1960s in Mauritania,
and smaller ones subsequently in Senegal and Gambia.
It is still not known whether the species undergoes any
transatlantic migration or whether the colonies are entirely
separate. Colonies are often large, with up to several
thousand pairs. The young are looked after by their parents
for an unusually long time. Adults have been observed still
feeding chicks that are 5 months old.

FACT FILE

RANGE
Breeds on coasts of Central
America and W Africa;
winters on coasts to S

HABITAT
Coastal and marine

SIZE
18–21 in (46–53 cm)

SOOTY TERN
Onychoprion fuscatus

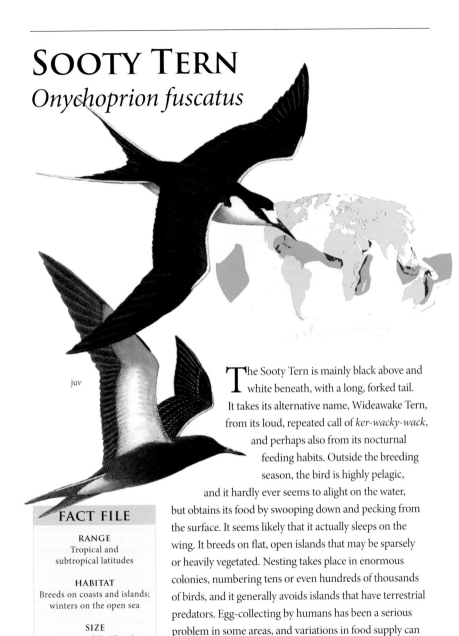

juv

FACT FILE

RANGE
Tropical and
subtropical latitudes

HABITAT
Breeds on coasts and islands;
winters on the open sea

SIZE
17–18 in (43–45 cm)

The Sooty Tern is mainly black above and white beneath, with a long, forked tail. It takes its alternative name, Wideawake Tern, from its loud, repeated call of *ker-wacky-wack*, and perhaps also from its nocturnal feeding habits. Outside the breeding season, the bird is highly pelagic, and it hardly ever seems to alight on the water, but obtains its food by swooping down and pecking from the surface. It seems likely that it actually sleeps on the wing. It breeds on flat, open islands that may be sparsely or heavily vegetated. Nesting takes place in enormous colonies, numbering tens or even hundreds of thousands of birds, and it generally avoids islands that have terrestrial predators. Egg-collecting by humans has been a serious problem in some areas, and variations in food supply can also cause significant fluctuations in breeding success.

BLACK SKIMMER
Rynchops niger

Like the other 2 species of skimmers, the Black Skimmer has an extraordinary bill in which the lower mandible is flattened and is a third as long again as the rounded upper mandible. When feeding, the bird flies close to the surface of the water, with only the tip of the lower mandible submerged. As soon as the lower mandible encounters prey, such as small fish and shrimp, the bird throws its head downward and snaps its bill shut. Much feeding takes place at dawn and dusk, and also by moonlight. Uniquely among birds, skimmers have vertical pupils in their eyes that can be closed to slits in bright sunshine, or opened to full circles in poor light.

FACT FILE

RANGE
S North America, Caribbean, and South America

HABITAT
Coastal and riverine marshes

SIZE
16–19½ in (40–50 cm)

Top 100 U.S. Wetland Birds and Seabirds

American Avocet
Size: 17–18½ in
Diet: Aquatic invertebrates
Range: SW U.S.A., coastal parts of Florida; Mexico, Caribbean

American Bittern
Size: 25 in
Diet: Fish
Range: C Canada to C U.S.A., S U.S.A., Mexico, Central America

American Black Duck
Size: 21–24 in
Diet: Seeds, roots, stems, grain, aquatic plants and insects, crustaceans, mollusks, some fish
Range: NE North America, S to Gulf of Mexico

American Coot
Size: 16 in
Diet: Plants
Range: S Canada, U.S.A., Central America, Caribbean

American Oystercatcher
Size: 16–17½ in
Diet: Insects
Range: Coastal parts of S U.S.A., Florida, coastal parts of Central America, coastal parts of W and E South America; Caribbean

American White Pelican
Size: 50–65 in
Diet: Fish
Range: U.S.A., Mexico

American Wigeon
Size: 18–22 in
Diet: Aquatic plants; insects, mollusks during breeding season
Range: N and C North America, S to Gulf of Mexico

Arctic Loon
Size: 23–29 in
Diet: Fish
Range: Arctic, N North America

Baird's Sandpiper
Size: 5½–7 in
Diet: Insects
Range: Arctic, Alaska, N to S Canada, N to S U.S.A., Central America, SW South America

Belted Kingfisher
Size: 12 in
Diet: Aquatic invertebrates, fish, insects, small vertebrates
Range: North America, S to West Indies and N South America

Black-bellied Plover
Size: 10½–12 in
Diet: Insects on breeding grounds, invertebrates, bivalves, crustaceans on wintering grounds

Range: Arctic, coastal parts of North America and South America

Black-crowned Night Heron
Size: 23–25½ in
Diet: Aquatic invertebrates, fish, amphibians, lizards, snakes, rodents, eggs
Range: North America, Central America, South America, Caribbean

Black-necked Stilt
Size: 14–15½ in
Diet: Insects
Range: S U.S.A., Central America, N and E South America, Caribbean

Black Scoter
Size: 17–21 in
Diet: Aquatic invertebrates, some vegetation
Range: Arctic, N Canada

Black Tern
Size: 8½–9½ in
Diet: Insects
Range: C North America

Blue-winged Teal
Size: 14–16 in
Diet: Aquatic invertebrates, seeds, plants
Range: S Canada, U.S.A., Central America, N South America, Caribbean

Bonaparte's Gull
Size: 11–15 in
Diet: Small fish, large invertebrates including insects
Range: Alaska, SW Canada, U.S.A., Caribbean

Brandt's Cormorant
Size: 27½–31 in
Diet: Fish, some squid
Range: Coastal parts of W Canada and U.S.A.

Brant
Size: 22–26 in
Diet: Plants
Range: Arctic, coastal parts of Alaska; N Canada, coastal parts of W Canada and W and E U.S.A.

Brown Pelican
Size: 43–54 in
Diet: Fish, some marine invertebrates
Range: Coastal parts of North America and South America, Galapagos Islands

Bufflehead
Size: 12½–16 in
Diet: Insects, crustaceans, mollusks, some seeds
Range: Alaska, Canada, U.S.A., Caribbean

Canada Goose
Size: 21½–43 in
Diet: Grasses and sedges in spring and summer, berries and seeds in fall and winter
Range: Arctic, North America

Canvasback
Size: 19–24 in
Diet: Seeds, buds, leaves, tubers, roots, snails, insect larvae
Range: S Canada to Mexico

Cattle Egret
Size: 19–21 in
Diet: Grasshoppers, crickets, spiders, flies, frogs, moths
Range: S U.S.A., N South America

Cinnamon Teal
Size: 14–17 in
Diet: Seeds, aquatic vegetation, aquatic and semi-terrestrial insects, snails, zooplankton
Range: W U.S.A., Mexico, N and SW South America

Common Goldeneye
Size: 16½–19½ in
Diet: Aquatic invertebrates, occasionally small fish, vegetation
Range: North America

Common Loon
Size: 27–36 in
Diet: Fish, some other aquatic vertebrates and invertebrates
Range: N North America

Common Merganser
Size: 21½–28 in
Diet: Small fish, insects, mollusks, crustaceans, worms, frogs, small mammals, birds, plants
Range: S Alaska and Canada, N U.S.A.

Common Moorhen
Size: 14 in
Diet: Seeds of grasses and sedges, some snails
Range: North America, South America

Common Tern
Size: 12–15 in
Diet: Small fish, some invertebrates
Range: S Canada, N U.S.A. to Florida; coastal parts of South America; Caribbean

Double-crested Cormorant
Size: 27½–35½ in
Diet: Fish, some other aquatic animals, insects, amphibians
Range: Coastal parts of S Alaska and W Canada; S Canada, U.S.A.

Dunlin
Size: 6–8½ in
Diet: Insects
Range: Arctic, parts of Canada, U.S.A.

Forster's Tern
Size: 13–14 in
Diet: Small fish and arthropods
Range: S Canada, U.S.A., coastal parts of Central America, Caribbean

Franklin's Gull
Size: 12½–14 in
Diet: Insects, earthworms, fish, mice, garbage, seeds
Range: S Canada, N U.S.A. to W Central America, coastal parts of W South America

Great Blue Heron
Size: 40–50 in
Diet: Fish, invertebrates, amphibians, reptiles, birds, small mammals
Range: S Canada, U.S.A., Central America, N South America, Caribbean

Great Cormorant
Size: 31½–39 in
Diet: Fish
Range: Coastal parts of E North America

Great Egret
Size: 37–41 in
Diet: Fish, invertebrates, amphibians, reptiles, birds, small mammals
Range: U.S.A., Central America, South America, Caribbean

Greater White-fronted Goose
Size: 25–32 in
Diet: Seeds, grain, grasses, sedges, berries
Range: Alaska, N Canada, W and S U.S.A.

Greater Yellowlegs
Size: 14 in
Diet: Small aquatic and terrestrial invertebrates, small fish, frogs, occasionally seeds, berries

Range: N North America to S South America

Green-backed Heron
Size: 16–19 in
Diet: Fish, frogs, snails, tadpoles, water lizards, crabs
Range: U.S.A., South America

Green-winged Teal
Size: 13–15 in
Diet: Seeds
Range: North America, Central America, Caribbean

Herring Gull
Size: 22–26 in
Diet: Fish, marine invertebrates, insects, birds, eggs, carrion, garbage
Range: Canada, U.S.A.

Hooded Merganser
Size: 16½–19½ in
Diet: Fish, aquatic insects, crustaceans (especially crayfish)
Range: S Alaska, Canada, U.S.A., Mexico

Killdeer
Size: 8–10 in
Diet: Invertebrates, also seeds, frogs, dead minnows
Range: S Canada to SC Mexico, N South America, West Indies

Laughing Gull
Size: 15½–18 in
Diet: Aquatic and terrestrial invertebrates, fish, squid, garbage, flying insects, berries

Range: Coastal parts of E and S U.S.A.; Florida, Central America, coastal parts of NW South America, coastal parts of Caribbean

Least Bittern
Size: 11–14 in
Diet: Small fish, insects
Range: W, E, and S U.S.A., Central America, parts of South America, Caribbean

Least Grebe
Size: 9–10½ in
Diet: Aquatic insects, small fish, tadpoles
Range: Parts of Central America and South America; Caribbean

Least Sandpiper
Size: 5–6 in
Diet: Insects
Range: Alaska, Canada, U.S.A., Central America, N South America, Caribbean

Least Tern
Size: 8–9 in
Diet: Small fish, some invertebrates
Range: Coastal parts of S U.S.A., coastal parts of N South America; Caribbean

Lesser Scaup
Size: 15½–18 in
Diet: Clams, snails, crustaceans, aquatic insects and plants
Range: C Alaska, NW and C Canada, U.S.A., coastal parts of N South America; Caribbean

Lesser Yellowlegs
Size: 9–10 in
Diet: Aquatic and terrestrial invertebrates (especially flies, beetles), occasionally small fish, seeds
Range: Alaska, NW and C Canada, U.S.A., Central America, South America, Caribbean

Little Blue Heron
Size: 22–29 in
Diet: Small fish, amphibians, aquatic invertebrates
Range: S U.S.A., Central America, N South America, coastal parts of W and E South America; Caribbean

Long-billed Curlew
Size: 20–25½ in
Diet: Insects
Range: C U.S.A., coastal parts of W and S U.S.A., Mexico

Long-billed Dowitcher
Size: 11½ in
Diet: Aquatic invertebrates, insects
Range: N Alaska, N Canada, W, E, and S U.S.A., Mexico

Mallard
Size: 20–25½ in
Diet: Insects, larvae, aquatic invertebrates and vegetation, seeds, acorns, grain
Range: North America

Manx Shearwater
Size: 12–15 in
Diet: Small fish, squid, crustaceans
Range: N Atlantic to South America

Marbled Godwit
Size: 16½–19 in
Diet: Insects
Range: S Canada, N U.S.A., coastal parts of S U.S.A., coastal parts of Central America

Mute Swan
Size: 49–61 in
Diet: Aquatic plants, some aquatic animals
Range: Parts of North America

Northern Fulmar
Size: 18–19½ in
Diet: Fish, squid, zooplankton, offal from fishing and whaling vessels, other animal matter found at sea
Range: Coastal parts of Canada

Northern Pintail
Size: 25–29 in (male), 17–25 in (female)
Diet: Grain, seeds, weeds, aquatic insects, crustaceans, snails
Range: North America, Central America

Northern Shoveler
Size: 17–20 in
Diet: Small swimming invertebrates, some seeds
Range: Arctic, North America, N South America

Osprey
Size: 21½–23 in
Diet: Fish
Range: North America

Parasitic Jaeger
Size: 17 in
Diet: Birds, small mammals, insects during summer, fish during winter
Range: N U.S.A.

Pectoral Sandpiper
Size: 9 in
Diet: Insects
Range: Arctic, N Alaska, NE to S Canada, NE to S U.S.A., Central America, N and C South America, Caribbean

Pelagic Cormorant
Size: 20–30 in
Diet: Fish, marine invertebrates
Range: Coastal parts of Alaska, W Canada and U.S.A.

Pied-billed Grebe
Size: 12–15 in
Diet: Fish, crustaceans (especially crayfish, aquatic insects)
Range: N to S Canada, S South America

Purple Gallinule
Size: 14 in
Diet: Seeds, flowers, fruit, grain, some invertebrates
Range: S U.S.A., Central America, NE South America, Caribbean

Red-breasted Merganser
Size: 20½–23 in
Diet: Fish, also crustaceans, insects, tadpoles
Range: North America, Gulf of Mexico

Reddish Egret
Size: 27½–31½ in
Diet: Fish
Range: Coastal parts of S U.S.A., coastal parts of Central America; Caribbean

Redhead
Size: 16½–21½ in
Diet: Plants
Range: E Alaska, S Canada, U.S.A., Mexico, Caribbean

Red Knot
Size: 9–10 in
Diet: Invertebrates (especially bivalves), small snails, crustaceans, also eats terrestrial invertebrates in breeding season
Range: Arctic Canada, South America

Red-throated Loon
Size: 21–27 in
Diet: Marine and freshwater fish
Range: Arctic, North America

Ring-billed Gull
Size: 17–21½ in
Diet: Fish, insects, rodents earthworms, grain, garbage
Range: S Canada, U.S.A., Mexico, Caribbean

Ring-necked Duck
Size: 15½–18 in
Diet: Plants

Range: S Canada, U.S.A., Mexico, S Central America, Caribbean

Ruddy Duck
Size: 14–17 in
Diet: Insects
Range: North America and South America

Ruddy Turnstone
Size: 8–10 in
Diet: Aquatic invertebrates and insects, carrion, garbage, bird's eggs
Range: Arctic, parts of North America and South America

Sanderling
Size: 7–8 in
Diet: Aquatic and terrestrial invertebrates
Range: Arctic, coastal parts of Alaska, Canada, U.S.A., Central America, and South America; Caribbean

Semipalmated Plover
Size: 7–7½ in
Diet: Insects
Range: Alaska, Canada, U.S.A., coastal parts of Central America and South America; Caribbean

Semipalmated Sandpiper
Size: 5–6 in
Diet: Insects
Range: N Alaska, N to S Canada, NE to S U.S.A., coastal parts of Central America, coastal parts of N South America; Caribbean

Short-billed Dowitcher
Size: 10–11½ in
Diet: Aquatic invertebrates, fly larvae, other insects, snails, some seeds on breeding grounds
Range: W, C, and E Canada, coastal parts of U.S.A. and Central America, coastal parts of N South America; Caribbean

Snow Goose
Size: 25½–33 in
Diet: Variety of plant species and parts, from aquatic plants to grasses and grain
Range: Arctic North America, S to Gulf of Mexico

Snowy Egret
Size: 22–26 in
Diet: Insects
Range: U.S.A., Central America, South America, Caribbean

Solitary Sandpiper
Size: 7½–9 in
Diet: Insects
Range: Alaska, S Canada, NE to S U.S.A., Central America, N to C South America, Caribbean

Spotted Sandpiper
Size: 7½ in
Diet: Aquatic and terrestrial invertebrates
Range: North America, Mexico to South America

Stilt Sandpiper
Size: 8–9 in
Diet: Insects
Range: N Alaska, N and S Canada, NE to S U.S.A., Central America, N and C South America, Caribbean

Tricolored Heron
Size: 23½–27½ in
Diet: Fish
Range: Coastal parts of S U.S.A., Central America, coastal parts of N South America, Caribbean

Trumpeter Swan
Size: 54½–62 in
Diet: Submerged and emergent aquatic vegetation, grasses, grain
Range: Parts of Alaska and Canada

Tundra Swan
Size: 47–59 in
Diet: Aquatic plants, seeds, tubers, grain, some mollusks and arthropods
Range: Arctic North America to S

Western Grebe
Size: 22–29 in
Diet: Fish
Range: W North America, from S Canada to Mexico

Western Sandpiper
Size: 5½–7 in
Diet: Insects
Range: NW Alaska, S U.S.A., coastal parts of N South America

Whimbrel
Size: 17½ in
Diet: Marine invertebrates, also insects, berries, flowers
Range: Parts of Alaska, and Canada, coastal parts of Canada, U.S.A., Central America, and South America; Caribbean

White-faced Ibis
Size: 18–22 in
Diet: Insects
Range: C and S U.S.A., Mexico, SE South America

White Ibis
Size: 22–27 in
Diet: Insects
Range: Coastal parts of S U.S.A. and Central America, Florida, N South America, Caribbean

White-rumped Sandpiper
Size: 6–7 in
Diet: Insects
Range: Arctic, N Alaska, E Canada and U.S.A. to S U.S.A., E Central America, South America, Caribbean

Willet
Size: 13–16 in
Diet: Insects
Range: S Canada, N U.S.A., coastal parts of U.S.A., Central America, and N South America, Caribbean

Wilson's Phalarope
Size: 9 in
Diet: Small aquatic invertebrates
Range: C North America, W and S South America

Wilson's Snipe
Size: 10½–12½ in
Diet: Larval insects, worms, crustaceans, mollusks, some vegetation and seeds
Range: Alaska, Canada, U.S.A., Central America, N South America, Caribbean

Wood Duck
Size: 17–20 in
Diet: Seed, acorns, fruit, aquatic and terrestrial invertebrates
Range: North America

Wood Stork
Size: 34–40 in
Diet: Fish
Range: SE U.S.A., Mexico, Central America, W to C South America

Yellow-crowned Night Heron
Size: 22–27½ in
Diet: Insects
Range: S U.S.A., coastal parts of Central America, N South America, coastal parts of E South America; Caribbean

GLOSSARY

abrasion Wear and tear on feathers, often removing paler spots and fringes and fading darker colors.

albinism A lack of pigment. True albinos are white with pink eyes, but most "white" birds are partial albinos, or albinistic, with patches of white and normal eye colors.

axillaries The feathers under the base of the wing, in the "wingpit." Also known as axillars.

band A metal band placed around a bird's leg, with an individual number; when the bird is caught or found dead, its movements can be traced. Also known as a ring.

beak Synonymous with bill; the two jaws and their horny covering.

bird of prey Usually refers to daytime birds of prey, including eagles, vultures, hawks, falcons, harriers, and kites; may be used to include owls. Also called "raptors," or raptorial birds.

brood A set of young birds hatched from one clutch of eggs.

call note A vocalization, usually characteristic of the species, made to maintain contact, warn of danger, or for other specific purposes.

cap A patch of color on the top of a bird's head, usually on the feathers of the forehead and crown.

carpal joint The bend of the wing, at the "wrist."

chick A young bird before it is able to fly.

clutch A set of eggs laid and incubated together in the nest; some species have several clutches during one breeding season, others ("single brooded") have only one.

colony A group of nests close together, often on the ground or in trees.

color ring or band A plastic or metal band placed on a bird's leg; a combination of colors or numbers on the band allow individual recognition without having to capture the bird.

corvid A bird of the crow family or corvidae.

courtship Usually ritualized behavior—male and female together forming a pair bond before breeding.

cryptic Describes coloration that gives a bird camouflage or makes it harder to see.

dawn chorus The loud chorus of bird song heard in spring from just before dawn.

display A form of ritualized behavior with a specific function, for example in courtship, or in distracting potential predators.

distribution The geographical range of a species, often split into breeding range, wintering range, and areas in which it may be seen on migration.

drake A male duck (females are then "ducks").

drumming The sound made in spring by a woodpecker vibrating its bill against a branch; also made by a snipe diving through the air with outer tail feathers extended and vibrating.

dusting "Bathing" in loose, dry sand, dust, or soil to help remove parasites from feathers.

eclipse A dull plumage worn by male ducks and geese in summer.

extinct Describes a species no longer living anywhere on Earth. If a species has disappeared from a country or region, but is still found elsewhere, it is properly described as having been "extirpated" from that area.

fall A sudden large arrival of migrant birds, especially when caused by bad weather on the coast.

feral Describes a bird or species that has escaped from captivity to live wild.

field "In the field" means "in the wild" or out of doors (as opposed to being captive, or held "in the hand").

field guide An identification guide to birds as they are seen wild and free.

fledgling A young bird that has just learned to fly and has its first covering of feathers.

flock A group of birds behaving in some sort of unison. Tight flocks (e.g., starlings in flight) are obvious, but loose, feeding flocks of birds in woodland may be less so.

game bird Bird commonly shot for sport—usually used to describe one of the pheasant, partridge, grouse, or quail families. Other birds include ducks and geese ("waterfowl").

genus A category in classification, above species, indicating close relationships. Appears as the first word in a two- or three-word scientific name. Plural is "genera."

gorget Band of color or pattern, such as streaks, around the bird's upper breast.

habitat The environment that a species requires for survival. Its characteristics include shelter, water, food, feeding areas, nest sites, and roosting sites. More loosely described in such terms as "lowland heath" or "deciduous woodland;" also used for particular times of year or types of behavior, e.g., muddy estuary, open sea, ploughed fields.

hen A female bird.

immature Describes a bird not yet old enough to breed or have full adult plumage colors.

incubation Maintenance of proper temperature of the egg to allow development of the embryo.

juvenile The young bird in its first full plumage. Also known as juvenal in the U.S.

loafing Sitting or standing, often in groups, apparently doing little or nothing. Gulls, for example, "loaf" for hours at a time.

mandible The jaw and its horny sheath; upper and lower mandibles together form the beak or bill.

measurements The size of a bird is usually indicated by the length from bill tip to tail tip on a bird laid out on a flat surface. In reality, the "size" depends as much on shape and bulk as on length.

migration A regular, seasonal movement of birds from one region or continent to another, between alternate areas occupied at different times of year.

molt The replacement of a bird's feathers, in a regular sequence characteristic of each species. There may be a complete molt or a partial molt depending on the season.

nest A receptacle built to take a clutch of eggs and, in many species, the young birds before they are able to fly; eggs may also be laid on a bare ledge or on the ground, with no nest structure being made.

nocturnal Active at night.

numbers Bird populations vary hugely from season to season, so are best described in terms of a particular measure that is easily repeated, usually "breeding pairs." In the case of large, more easily counted birds, such as ducks and geese, the measure is the total number of individuals at a certain season.

ornithology The study of birds. Usually refers to scientific study of biology and ecology, while the hobby of watching birds is known simply as bird-watching or birding.

passage migrant A species or bird seen in some intermediate area during its migration from summer to winter quarters (or vice versa).

passerine A "perching bird."

plumage A covering of feathers; also often used to describe the overall colors and patterns of the feathers, defining a bird's appearance according to age, sex, and season.

preening Care of the feathers, especially using the bill to "zip" the structures back into place.

race A recognizable geographical group, or subspecies, within a species. Often there is no obvious border between groups, which blend (in a "cline") from one

extreme to another. There may be more distinctive differences between isolated areas, such as islands, in which case the decision whether there are races, or separate species, can be difficult.

rarity An individual bird in an area where it is not normally seen, or is seen in only very small numbers. A species with a small world population is "rare."

roost To sleep; also the area where birds sleep.

seabird A species that comes to land to nest, but otherwise lives at sea and is not normally seen inland.

shorebird A plover, sandpiper, or related species; outside of North America, usually called a "wader." Neither word is entirely satisfactory.

soaring Flight, often at a high level, in which the wings are held almost still, using air currents for lift and propulsion.

song A vocalization with a specific purpose and usually distinctive for each species. In particular, advertising the presence of a bird on its territory.

species A group, or groups, of individuals that can produce fertile young. Different species rarely interbreed naturally; if they do so, infertile hybrid offspring are produced.

territory An area defended for exclusive use by an individual bird or a family. Both breeding and winter-feeding territories may be defended.

waterfowl Ducks, geese, and swans. Also known as wildfowl.

INDEX

Acknowledgments

Illustrations Norman Arlott, Hilary Burn, Chris Christoforou, Robert Gillmor, Peter Hayman, Denys Ovenden, David Quinn, Andrew Robinson, Chris Rose, Ken Wood, Michael Woods